CELESTIAL NAVIGATION

Use of sextant by Harvard student in navigation. Photo: Stephen G. Perrin.

CELESTIAL NAVIGATION

SECOND EDITION

Frances W. Wright

CORNELL MARITIME PRESS

Centreville, Maryland

Library of Congress Cataloging-in-Publication Data

Wright, Frances Woodworth.
  Celestial navigation.

  Bibliography: p.
  Includes index.
  1. Navigation. 2. Nautical astronomy. I. Title.
VK555.W9   1982   623.89   82-4964
ISBN 0-87033-291-0   AACR2

Manufactured in the United States of America
First edition, 1969. Second edition, 1982; fifth printing, 1998

Dedicated to Students of Astro. 2, Harvard

"I had to put my faith in the sextant."

Sir Francis Chichester
The Lonely Sea and the Sky

Contents

Acknowledgments

In acknowledging aid from many others, I appreciate most of all the stimulus, enthusiasm, and inspiration given to me by the students of Astro. 2 and 91r at Harvard.

Several scientists and mariners critically read the whole manuscript and contributed helpful comments and suggestions. I am especially grateful to John R. Burke and James Elliot, to Dr. Paul W. Hodge of the University of Washington, to Palle Mogensen of the Arctic Institute of North America, and to Dr. Carl S. Nilsson, University of Adelaide, Australia. Harold Boeschenstein, Jr., Dr. Fred A. Franklin, and Nathan Hazen, Harvard, also rendered valuable assistance. I am indebted to Solomon Conn for the sketch in Part II, and I thank Walter M. Hunkeler, who appears in the frontispiece. Stanley R. Larson drew the figures.

I appreciate the support of Dr. Bradford Washburn, Director of Boston's Museum of Science, and of John M. Carr, Director of the Hayden Planetarium (Boston), who invited me to give the lectures upon which this manuscript is based.

On the S. S. *Independence*, in August, 1967, the following officers offered me all possible courtesy and aid in practicing the techniques of celestial navigation on the bridge: Charles E. Reilly, Master, Robert Adams, Staff Captain, Anthony Palazzolo, Chief Officer, Robert J. Thorpe, Second Officer, Robert P. Ezell, Second Officer, and Joseph Bunicci, Third Officer.

I wish to acknowledge the cooperation extended by the heads of several government agencies, especially by Dr. Raynor L. Duncombe, Director of the Nautical Almanac Office, by the Commander of the U. S. Naval Oceanographic Office, by George N. Weston, staff scientist of the U. S. Naval Oceanographic Office, by Fred L. Whipple, Director of the Smithsonian Astrophysical Observatory, and George Field, Director of the Center for Astrophysics. I also appreciate receiving from Leroy E. Doggett, two advance copies of August daily pages of the 1983 edition of *The Nautical Almanac*, and I am indebted to Ernest B. Brown of the Defense Mapping Agency Hydrographic Center for help with various items of change and improvement.

I thank the following authors for the stimulus communicated to this manuscript by their writings: J. H. Blythe, Sir Francis Chichester, R. L. Duncombe, Rupert T. Gould (R. N.), Richard Hough, Alfred Lansing, Alvin Moscow, M. W. Richey, Larry Rockefeller, D. H. Sadler, and Eric Tabarly.

Finally, I shall always feel special gratitude for the inspiration and encouragement received from Professor Bart J. Bok, coauthor of our *Basic Marine Navigation*. I value help from Dr. John J. Black and Tom Francis, and I appreciate all the criticism and thorough checking of this second edition by William M. Blair, Brian P. Cooke, and Mark H. Helmericks, three of my Harvard navigation students.

The Parts

This guide to celestial navigation is divided into parts that are somewhat independent, to satisfy the various needs of users, depending upon their experience and requirements. Parts I, II, III, and most of V and VI provide the background theory for an understanding of all the basic principles of celestial navigation. These five parts might be omitted by a navigator or yachtsman who wishes only to refresh his knowledge of celestial navigation, and who wants to become familiar with the use of *Pub. No. 249 Sight Reduction Tables for Air Navigation* by working through the reductions in the practice cruise of the S. S. *Ruchbah* in Part VII.

Part IV gives two techniques for the determination of latitude at meridian passage of the sun. This part also shows how an approximate longitude may be obtained by a "noon curve." Part V includes an explanation of a special technique originated by Sir Francis Chichester. Part VI, besides describing different kinds of solar time,

shows details, in Section W, of using *Pub. No. 249* to give a fairly accurate time at a known position. Part VII presents the practice cruise of a day at sea with celestial navigation. Part VIII contains supplementary explanation of details in the practice cruise of Part VII. Part IX explains the application of celestial navigation as a check of the magnetic compass on any course. Part X contains the necessary extracts from the *Nautical Almanac* and from *Pub. No. 249 Sight Reduction Tables* for reduction of sextant sights of the practice cruise. Filled-in forms, with details of answers to all problems of the practice cruise, are included. Extra forms, for future navigation, are added. Part XI lists abbreviations and symbols, as well as references.

Some knowledge of plotting and coastwise navigation has been assumed. The author's *Coastwise Navigation* or Bowditch *Pub. No. 9 American Practical Navigator* may be used for reference.

Foreword

This book has grown out of my many years of experience in teaching Astronomy 2 (Navigation) at Harvard University. It is intended to answer the need, expressed by many of my friends and acquaintances, for a quick, easy, and thorough explanation, *with realistic worked examples*, of the practice of celestial navigation at sea, using simple and inexpensive equipment.

Navigation has not only been my profession for years while I taught it, but has also been the source of great enjoyment as my hobby. I have navigated across the Atlantic four times, and on the Caribbean and the Pacific, always on passenger liners, working usually from the deck, in a sort of ex-officio, normally uninvited competition with the regular ship's navigators. I used the methods described in this book, methods understandable to anyone and available for use with only a simple sextant, watch, and some tables. One of these navigational cruises was on the S. S. *Independence*, from New York to Naples, in 1967. A practice cruise, Part VII, based on actual sextant sights I took during a day's navigating on the bridge of the *Independence*, was revised for 1983 in this edition.

I use in this book *Pub. No. 249*, a publication of the Defense Mapping Agency Hydrographic Center entitled *Sight Reduction Tables for Air Navigation*. Originally designed for air navigation, the tables can just as conveniently be used at sea. They are so quick and easy to use, especially Volume 1 for stars, that they have become very popular for marine as well as air navigation.

Much of the "theory" given in this book is based in format on the series of lectures I gave from 1967 to 1981 at the Hayden Planetarium in Boston's Museum of Science. Since the audience was made up of the general public, many of whom had virtually no background in either celestial navigation or astronomy, I made the treatment simple and basic. A few special navigational techniques, however, are added here for the use and interest of experienced navigators.

Since 1969 when the first edition was published, I have more than tripled the former number of oceanic cruises on passenger liners, and the same factor holds true with the number of series of lectures on celestial navigation at Boston's Museum of Science. In the 1970s three of my Harvard students sailed around the world on their own voyages, while a fourth sailed solo across the Atlantic from England. I am indebted to Andy Burnes, Bill Fitz, Brad Ives, and Phineas Sprague for showing that the techniques explained in this book—*Celestial Navigation*—along with constant vigilance, made these voyages successful.

It should be mentioned that the methods of this book can be used to advantage by anyone who has need to know the apparent positions of the sun, moon, or stars. Architects concerned with sunlight and shadow, or lawyers concerned with court cases involving these problems, for example, will find that the techniques described here are fast, simple, and accurate.

F. W. W.

CELESTIAL NAVIGATION

INTRODUCTION

Definition and Today's Use of Celestial Navigation

Navigation, in modern usage, is defined as keeping track of the past, present, and future position of a vessel. Celestial navigation, using the sextant, accurate timepiece, and *Nautical Almanac,* is still an extremely important method of navigation today on a yacht, ocean liner, or man-of-war. This may surprise many who have thought that automatic, electronic, and satellite navigation would replace older navigational systems. Although there have been great developments along these lines, today a successful navigator should know several techniques so that he can adopt the one most appropriate to the situation, or use one to check another technique, and the fact remains that celestial navigation is a fundamental procedure on most navigable craft. In the words of Blythe, Duncombe, and Sadler (U. S. Naval Oceanographic Office, U. S. Naval Observatory, and Royal Greenwich Observatory, respectively): "In spite of tremendous developments in navigational systems of all kinds, navigators are today still vitally interested in improved facilities for sight reduction, and in more rapid and accurate techniques for establishing the line of position." A navigator will certainly use electronic techniques when they are available and rules permit, but he can still perform alone, quite independently of electronics. If the latter should fail at any time, celestial navigation could save the navigator's life or any other lives involved; in many cases celestial is more accurate than electronic navigation.

As the name suggests, celestial navigation uses the celestial bodies as its most important tools; these bodies are the sun, moon, and planets of our solar system, and the stars in our immediate universe. Hence, these bodies are all universal and can be used universally by all persons on the earth at appropriate times. No fee is collected for a celestial sight, no permission is necessary, and not even a ticket is required, but a navigator does need a sextant, an accurate watch or other timepiece, a *Nautical Almanac,* and some simple tables for reducing his celestial sights. As for mathematics, the only requisite is an ability to add and subtract figures without making a mistake. The watchword for successful navigation is *constant vigilance.*

PART I

Descriptive Astronomy Necessary for an Understanding of the First Principles of Celestial Navigation

A
Definitions and Projection of Earth onto Celestial Sphere

CELESTIAL SPHERE. It is well to understand some of the theory behind the reductions which the navigator makes. Often the theory becomes clear if we are able, in imagination, to travel back and forth from the earth to the celestial sphere. This is a fictitious sphere which astronomers have invented. It is at an infinite distance, but since we really cannot represent infinity in a figure on a page, the reader must imagine that the outer circle, drawn in sketches to represent the celestial sphere, is actually at an *infinite* distance. Almost always, in celestial navigation, we can consider the earth as a perfect sphere at the center of the celestial sphere. The latter is a convenient background on which we can project ourselves and important points on the earth. The projection is from "C" in Fig. 1, the center of the celestial sphere and also the center of the earth.

ZENITH. If the navigator is on a ship at 0, for instance, his position 0 on the earth can be projected onto his zenith (Z). The zenith is the point directly overhead.

CELESTIAL POLES. The North Pole on the earth can be projected onto the north celestial pole; the South Pole onto the south celestial pole.

CELESTIAL EQUATOR. The earth's equator, if the plane of this equator is extended an infinite distance, can be projected onto the celestial equator, which is a very important great circle in celestial navigation.

GREAT CIRCLE. Any circle described on a sphere so that its plane passes through the center of the sphere.

DECLINATION. Similar to a celestial latitude, declination is measured in degrees, minutes, and seconds due north or south from the celestial equator. Declination varies from 0° on the celestial equator to 90° north or south at the celestial poles, depending upon whether the celestial body is at the north or the south celestial pole.

HOUR CIRCLES. Declination is measured along hour circles, which can be projected upon the meridians of the earth, or vice versa. In Fig. 1 the north celestial pole is 90° from the celestial equator, and has a declination of 90° N.

B
Horizon and Cardinal Points of Horizon

HORIZON. If the navigator looks out on the water as far as he can see, he is looking at the plane of his horizon, and if this plane is extended until it intersects the celestial sphere in a great circle, we call this great circle the navigator's celestial horizon, and mark the north point (N) of the horizon directly under the north celestial pole in the Northern Hemisphere. We mark the south point (S) 180° from N, in the opposite direction, and we mark the east and west points (E and W) exactly halfway between N and S. E is on the right, and W on the left as the navigator faces north.

CARDINAL POINTS OF HORIZON. Since we know from solid geometry that great circles bisect each other, E and W will also be the two intersections of the celestial equator and the navigator's horizon (Fig. 1). Points N, S, E, and W are called cardinal points of the navigator's horizon.

5

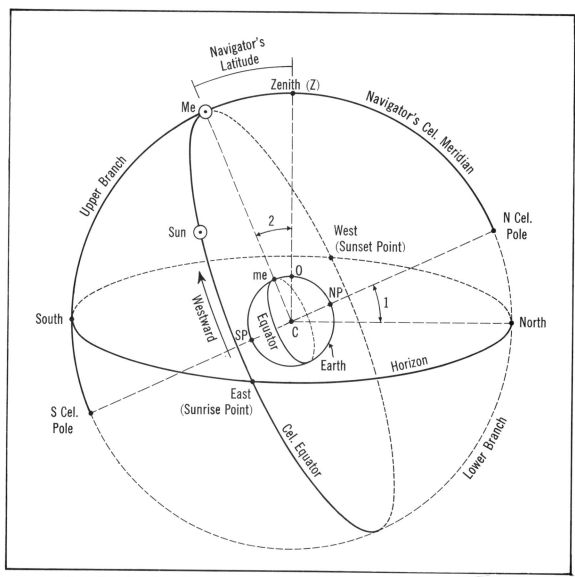

Fig. I. Celestial sphere, March 21, with earth at center (radius of celestial sphere is infinite).

C
Altitude, Zenith Distance, and Celestial Meridian

The Alt. (**altitude**) of a celestial body, by definition, is the number of degrees, minutes, and seconds a celestial body is above the navigator's horizon. **Vertical circle** is a great circle through the zenith and the celestial body on which the altitude is measured. On this same vertical circle, the distance in degrees of a celestial body from the navigator's zenith is by definition the ZD (**zenith distance**) of the celestial body. It follows, for a celestial body, that

$$\text{Alt.} + \text{ZD} = 90°,$$
or
$$\text{ZD} = 90° - \text{Alt.}$$

In Fig. 2, for example, SF = Alt. of sun

ZD of sun (SZ) = 90° − Alt. of sun (SF)

Another important great circle is the projection of the navigator's terrestrial meridian onto the navigator's celestial sphere.

The **celestial meridian** passes through the navigator's zenith, through the north and south celestial poles, and through the north and south points of the navigator's horizon. The celestial meridian has two branches, each extending from one celestial pole to the other.

UPPER AND LOWER BRANCHES OF ME-RIDIAN. The branch which runs from the celestial poles *through* the navigator's zenith is called the *upper branch* of the navigator's celestial merid-ian, and is the most important branch in celestial navigation. It is the black outside semicircle marked in Fig. 1. The broken semicircle marks the lower branch.

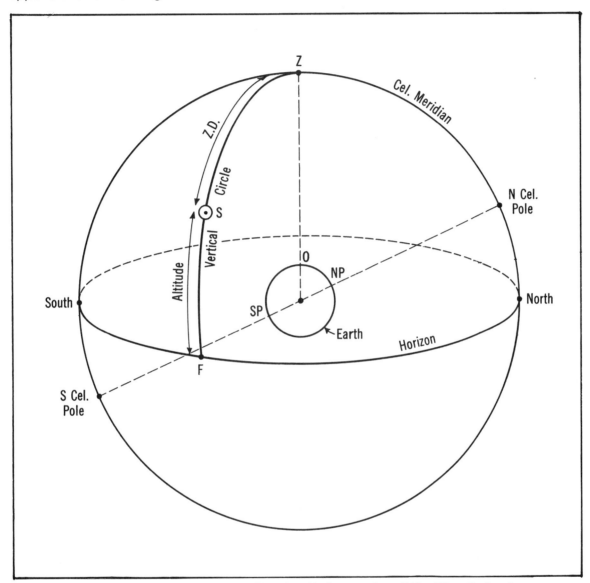

Fig. 2. Vertical circle through sun (S) and zenith distance of sun (90° − ZD = alt. of sun).

D
Diurnal Motions of the Sun; Ecliptic, Equinoxes, and Solstices

TWO APPARENT DIURNAL MOTIONS OF THE SUN. During the day, the sun appears to move westward and almost parallel to the celestial equator. This is called an apparent diurnal motion of the sun. It is due to the real eastward rotation of the earth on its axis in 24 hours. On March 21, for example, the sun rises practically due east (see Fig. 1). As the earth rotates east-ward that day, the celestial sphere seems to rotate westward, and the sun, projected onto the sphere, appears to move along the celestial equator westward. Its altitude at sunrise is 0°. As the sun apparently moves westward practically on the celestial equator during this entire day of March 21, it rises higher and higher in the sky—its altitude continually increases—until it reaches the upper branch of the celestial meridian at M_e in Fig. 1. At this point the sun's altitude is greatest

year, or through 360° in 365 days, the sun seems to move about 1° per day in this eastward motion.

ECLIPTIC. The sun's apparent path among the stars is in the plane of the earth's real orbit. This plane is called the plane of the ecliptic; the projection of this plane onto the celestial sphere

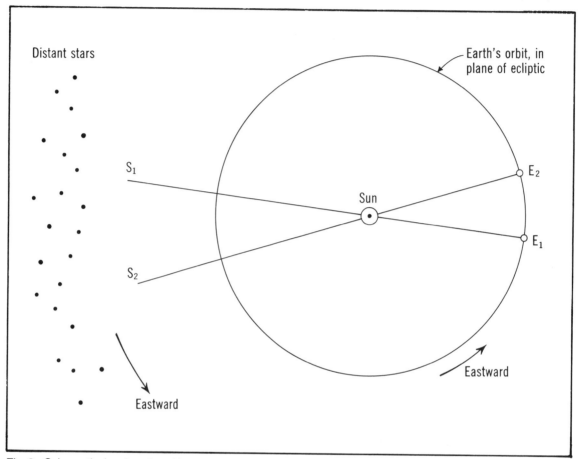

Fig. 3. Schematic drawing of earth's eastward path or orbit around sun (sun seems to move eastward among stars).

during that day, and the navigator can obtain his latitude from a noon sight (explained in detail below). Then, as the sun continues to move westward in this apparent motion, its altitude decreases until it reaches 0° at sunset.

The second apparent motion of the sun is eastward among the stars while the earth is actually revolving eastward around the sun once a year in a path, or orbit, as shown in Fig. 3. When the earth is at E_1, the sun seems to be at S_1 among the more distant stars. When the earth has moved eastward to E_2, the sun appears to be at S_2 among the stars, and so the sun has apparently moved eastward among the stars. Since the earth revolves all the way around the sun in a

is a great circle called the ecliptic. The sun, in its apparent diurnal motion eastward, seems to move 1° along the ecliptic each day. We should note, as shown in Fig. 4, that the great circle of the ecliptic intersects the great circle of the celestial equator in two important points.

VERNAL EQUINOX. One point is the first point of Aries, or the vernal equinox (expressed by the symbol ♈ in *Pub. No. 249*). The sun, near the center of the celestial sphere (sun not shown in Fig. 4) has this point for a background on March 21 (± 1 day) each year, afterwards traveling east and north of the celestial equator in its ecliptic-motion. The other point (the second of the two intersections of ecliptic and celestial equator) is the autumnal equinox, the point which the sun has for a background about September 23 each year. Due to these two apparent motions, the sun daily moves eastward 1° along the ecliptic, at the same time moving westward from sunrise to sunset in a diurnal circle which is practically parallel to the celestial equator. The sun's daily motion along the ecliptic can usually be ignored by the navigator, since the appropriate

DIURNAL CIRCLES OF THE SUN THROUGH THE YEAR. We are now ready to look at Fig. 5, in which the ecliptic is added to the drawing of the celestial sphere in Fig. 1. The ecliptic was added so that we might trace the diurnal circles of the sun (due to earth's rotation) through the year at lat. (latitude) 23½° north. The drawing is easier to understand if we assume a time when ♈ (the first point of Aries, and the intersection of ecliptic

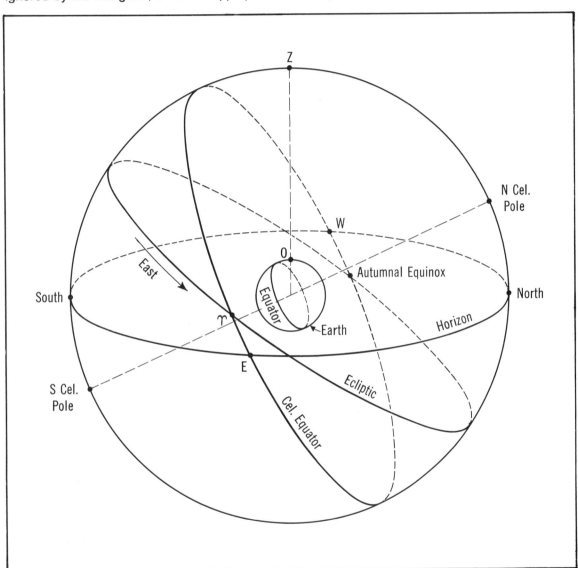

Fig. 4. Celestial sphere with ecliptic and equinoxes.

Nautical Almanac for the year gives the declination of the sun for every second of time during that year. The other motion is more important for the navigator as he uses this information when he takes a series of sextant sights before, during, and after meridian passage of the sun across his celestial meridian.

and celestial equator where ecliptic to eastward runs north of the equator) is just rising, due east. Again, let us suppose that the date is March 21. The sun will then have ♈ for a background. The sun will rise at E (due east) and travel along the celestial equator practically all day, setting at W. The next day, however, because of the sun's eastward motion along the ecliptic, it will rise about 1° north of E, and its daily circle will be approximately parallel to the Equator, but almost 1° farther north. The next day the sun's daily circle will be even farther north, and this will continue (as indicated by the faint curves north of the celestial equator in Fig. 5) until the time of the summer solstice, when the sun has traveled east along the ecliptic until it is as far from the celestial equator as the angle between the planes of the celestial equator and the ecliptic will permit.

terminology, but he remembers that when the sun reaches this point in the ecliptic, its distance from the celestial equator is 23½°, and its declination is then 23½° north, with the diurnal circle as drawn in Fig. 5. At this time the sun is "at the summer solstice." It crosses the celestial meridian at Z, the zenith in Fig. 5, because we assumed a latitude of 23½° north. When, however, the sun is "at the winter solstice," it has a declination of 23½° south, with its diurnal circle south of the celestial equator, as shown in Fig. 5. It crosses

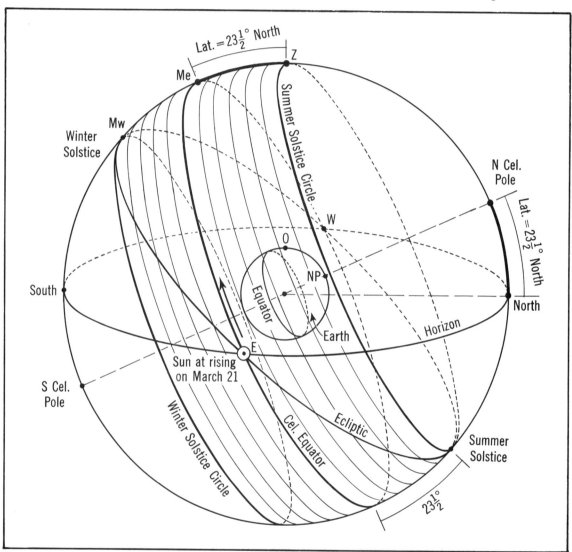

Fig. 5. Diurnal circles of the sun through the year at latitude 23½° north (circles are represented by the faint circles between the summer solstice circle and the winter solstice circle).

ANGLE BETWEEN ECLIPTIC AND CELESTIAL EQUATOR. This angle is 23½°, and is called by astronomers the obliquity of the ecliptic. The navigator does not have to worry about the

the celestial meridian at M$_w$.

At 70° north latitude, the diurnal circles of the sun will look as pictured in Fig. 6. It can be seen quite clearly from the illustration that near the time of the summer solstice at latitude 70° north, the sun never goes below the horizon, whereas at the time of the winter solstice the sun would be below the horizon during the whole 24 hours.

When, at *any* latitude the sun is close to the celestial equator, the number of hours of sunshine and darkness are nearly equal. This happens when the sun is "at one of the equinoxes."

E
Latitude Related to Zenith Distance of Celestial Equator: The First Basic Relation in Celestial Navigation

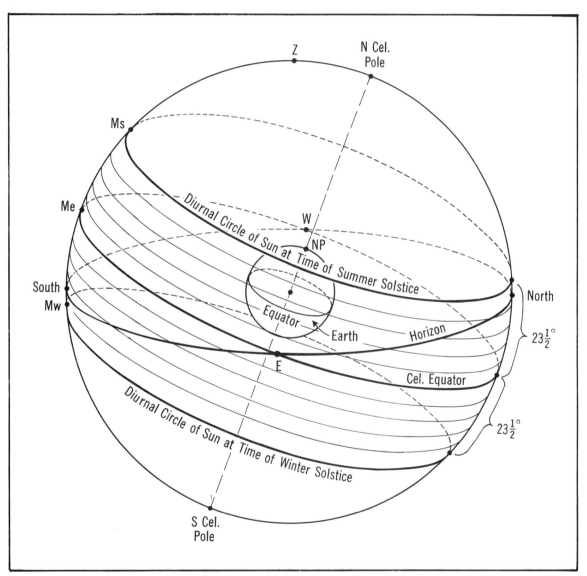

Fig. 6. Diurnal circles of the sun at 70° north latitude.

The First Basic Relation in Celestial Navigation states: *The zenith distance of the celestial equator equals the navigator's latitude.* This relation can be understood most easily by projecting upon the earth some points from the celestial sphere. For instance, if we bring the zenith distance of the celestial equator down upon the earth, the zenith, Z, in Fig. 1 falls upon the navigator's head at 0, and the celestial equator projects onto the earth's equator. In other words, the zenith distance of the celestial equator, ZM_e in Fig. 1, projects onto Om_e on the earth. Now we know that Om$_e$, the distance of the navigator from the earth's equator, is his latitude by definition.

Hence, by projection, we understand the above basic relation in celestial navigation.

F
Latitude Related to Altitude of Elevated Celestial Pole: The Second Basic Relation

The Second Basic Relation in Celestial Navigation states: *The altitude of the north celestial pole above the horizon equals the navigator's latitude in northern latitudes.* As seen in Fig. 1, on the celestial sphere the arc between the north point of the horizon and the north celestial pole, sub-

Therefore, we have shown the Second Basic Relation in Celestial Navigation as stated above. If the navigator should be at a southern latitude, the south celestial pole would be above the south point of the horizon, and again the altitude of the south celestial pole above the south point of the

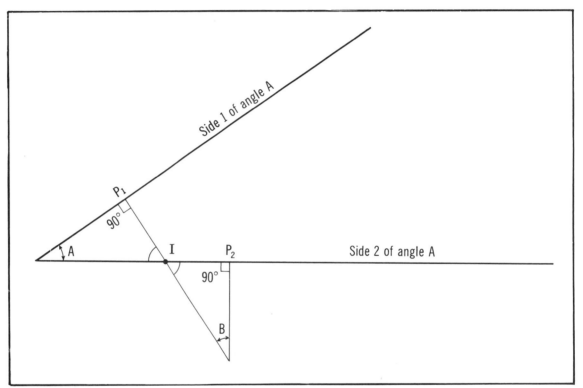

Fig. 7. Two angles with sides of one angle perpendicular to sides of other angle, respectively, are equal.

tended by the angle 1, equals the arc between the zenith and M$_e$, subtended by the angle marked 2, since angles with sides perpendicular, respectively, are equal. To understand this relation, in Fig. 7 draw BP$_1$ at right angles to side 1 of angle A. Draw BP$_2$ at right angles to side 2 of angle A. The vertical (opposite) angles at I are equal. The right angles of 90° are equal. The sum of the angles in any plane triangle equals 180°. Hence angle A equals angle B, or angles are equal if sides of one angle (A) are perpendicular, respectively, to sides of the other angle (B).

horizon equals the navigator's latitude (see Fig. 8). By simply estimating the altitude of the elevated celestial pole above the navigator's horizon, the navigator has some idea of his latitude. If he were at the equator, for example, with his latitude equal to 0°, he would find both celestial poles at an altitude of 0°, or lying on the horizon due north and due south, respectively. In another extreme case, at the North Pole, for instance, he would find the north celestial pole directly overhead, at his zenith, with an altitude of 90°.

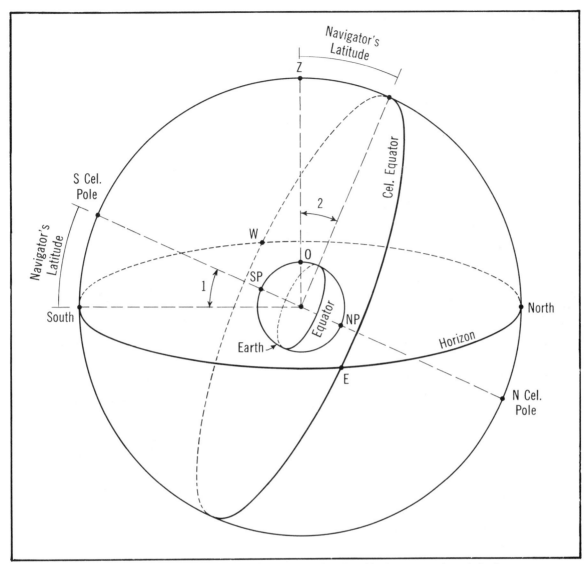

Fig. 8. South celestial pole elevated above south point of horizon at southern latitudes.

PART II

Identification of North and South Celestial Poles, Polaris, and Ruchbah

The navigator may wonder how he can recognize the north celestial pole if he lives in north latitude. The celestial pole has no neon light, nor a flag waving to mark its position.

POLARIS. One good easy method is to use Polaris, the North Star, which happens to be within

1° of the north celestial pole, which is marked by ⊕ in Fig. 9. But how does the navigator or the yachtsman find Polaris? That is the first problem, of the two, since Polaris is a star which can be seen only if we can identify it among thousands of other stars.

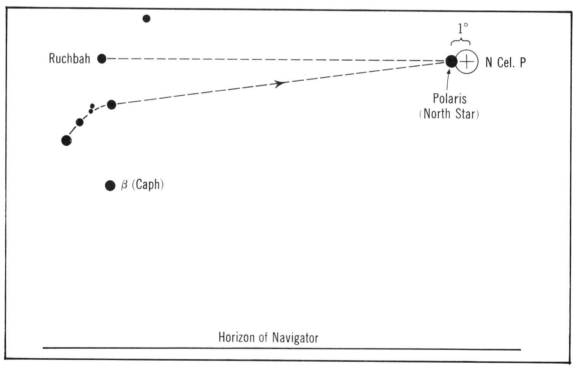

Fig. 9. CASSIOPEIA and curve of stars leading to Polaris (in this position altitude of Ruchbah equals altitude of north celestial pole).

G
Finding Polaris from CASSIOPEIA

If the navigator in mid-north latitudes can identify the constellation or group of stars called CAS-

SIOPEIA, he is taking the first step in the identification of Polaris. CASSIOPEIA, a queen on a throne

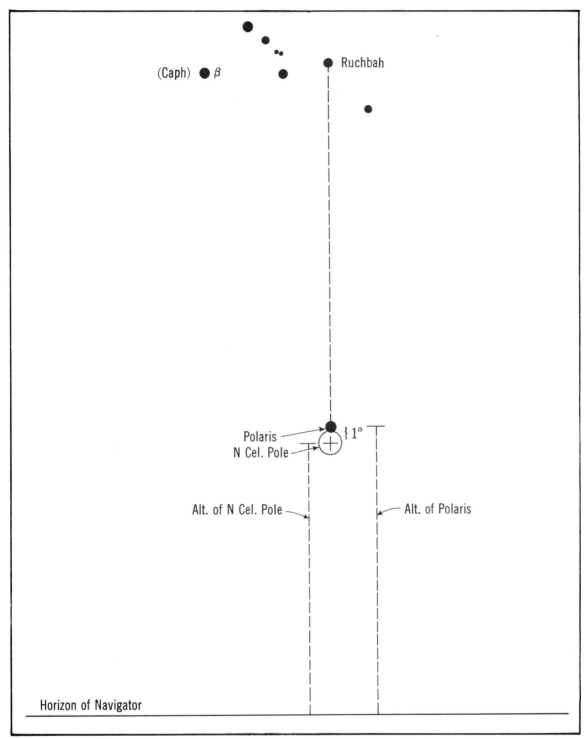

Fig. 10. Ruchbah above Polaris; 1° subtracted from altitude of Polaris gives altitude of north celestial pole, or the latitude north.

or regal chair, according to one story in ancient mythology, is a W-shaped arrangement of stars shown in Fig. 9. The navigator can then proceed in imaginative travel to Polaris and secondly to the north celestial pole. If the navigator follows

the curve, indicated in Fig. 9 of CASSIOPEIA, and traces this curve out to a distance equal to twice the distance across the W of CASSIOPEIA, he will identify and find the North Star, Polaris.

H

Use of Ruchbah in Finding North Celestial Pole and also in Determination of Approximate Latitude in Emergency

To find the celestial pole, next, it is helpful to use another star in CASSIOPEIA, by the name of Ruchbah, and labeled in Fig. 9. Ruchbah is of great aid to the navigator, especially in an emergency. If the navigator draws a straight line from Ruchbah

can measure with a sextant (Fig. 14) the altitude of Polaris above the horizon when he can see both Polaris and the horizon, as at evening and morning twilight. Then, by noting the position of the north celestial pole as indicated above, he

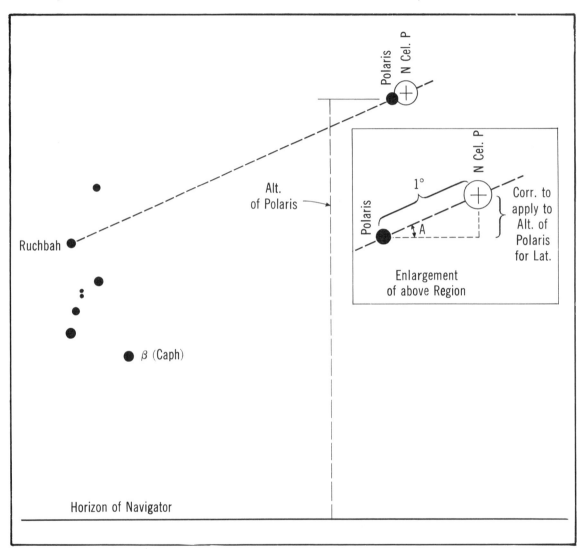

Fig. 11. Estimate of correction to apply to altitude of Polaris for latitude can be approximated visually (always a fraction of 1°).

to Polaris and then extends this straight line 1° to the *other* side of Polaris, he will find the position at that instant of the north celestial pole, marked by ⊕ in Fig. 9.

FINDING LATITUDE FROM POLARIS BY USE OF RUCHBAH TO GIVE NECESSARY CORREC-TION TO ALTITUDE OF POLARIS. The navigator

can estimate, even without a *Nautical Almanac,* the approximate altitude of the north celestial pole above his horizon. Thus, he obtains his latitude. If, for instance, the straight line drawn in imagination from Ruchbah to Polaris is parallel to the horizon (as in Fig. 9), the altitudes of Polaris and the north celestial pole will be equal, and there will be no correction to the altitude of Polaris, which will then give the navigator's latitude. If, on the other hand, CASSIOPEIA was above Polaris, and the straight line was at right angles to the horizon, as shown in Fig. 10, the correction of 1° would be subtracted from a sextant altitude of Polaris. In general, the approximate correction to be applied (added or subtracted) is the sine of the angle A in Fig. 11 (for those who are familiar with trigonometry), and with a little practice this correction can be estimated by the navigator in an emergency to within 10′ of arc, and hence to within 10 nautical miles. The navigator who does not use trigonometry can estimate this correction to within 20′, or 20 nautical miles, just by simple eye estimate. A very accurate correction can be found in the *Nautical Almanac* (pp. 274-76), with directions for use at the bottom of these pages.

"Wish my hands were not so cold, so that I could look in my Emergency Pamphlet and find out where I'm supposed to be with Polaris at the zenith." Sketch: Solomon Conn.

I
Finding True Azimuth from Polaris

Besides providing an estimate of latitude, Polaris has an even more important function for the navigator or yachtsman.

AZIMUTH FROM POLARIS. Polaris tells him, to within 1°, the direction of true north on his horizon. The navigator needs true north for orienting himself with respect to all true directions, which he uses in plotting his true course on a chart or in plotting true bearings or azimuths of celestial bodies. By definition, the azimuth of a celestial body is similar to the true bearing of a

light in piloting off the coast. It is the measure, at the navigator's position, of the angle between the direction to true north on the navigator's horizon and the direction to the point on the earth underneath the celestial body. This latter point is called the subsolar or substellar point, as the case may be (either for the sun or for the star used). The azimuth, just as any true bearing, is measured in a *clockwise* direction from the north, from the navigator's position.

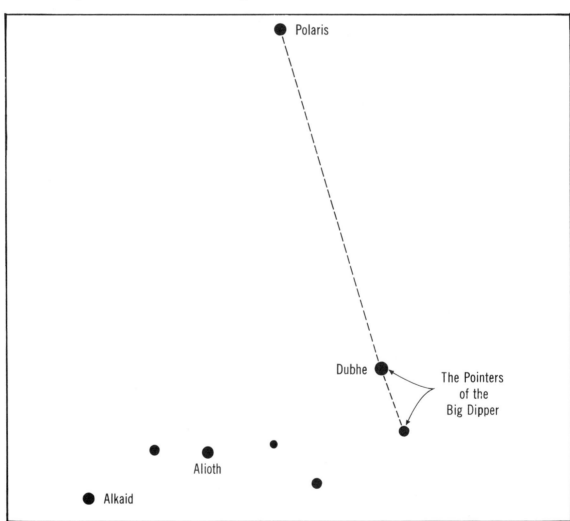

Fig. 12. Finding Polaris from Big Dipper of ᴜʀsᴀ ᴍᴀᴊᴏʀ.

J
Finding Polaris from the Big Dipper

Often, in the Northern Hemisphere, CASSIOPEIA will be low on the horizon, or practically invisible in all the haze hanging just above the horizon. It is then convenient to identify Polaris, the North Star, by using the Big Dipper in the constellation or group of stars called URSA MAJOR or the Great Bear. The handle of the dipper makes the tail of the bear. It is rather long for a bear's tail, but according to ancient mythology, the tail was stretched when the Great Bear was hurled into place in the sky by his tail. However, for a dipper, the handle is perfect, and the navigator finds the Dipper of more practical use to him than the Bear. If the navigator learns to recognize the Big Dipper, he can easily identify the pointers, the two endstars of the bowl of the Big Dipper (Fig. 12). If the Dipper is right side up for holding imaginary water, a line through the pointers above the bowl will lead to Polaris, the distance to Polaris from Dubhe, the "top" star of the point-

ers, being about 5 times the distance between the pointers. The distance between the pointers is about 5°, a convenient degree-stick for estimating other distances. It should be remembered that, in the lifetime of a navigator, the stars keep the same relations with respect to each other, and distances are maintained. For navigation on the earth the navigator may think of the stars as fixed points on the celestial sphere. Again it is the eastward rotation of the earth on its axis in 24 hours which causes the celestial sphere to have an apparent westward rotation of 360° during the whole day of 24 hours. The astronomer has to worry about individual motions of the stars, but the navigator on the earth may disregard these real motions of the stars if he adopts positions for all the celestial objects as given to him by the *Nautical Almanac* for the appropriate year, day, and time of day.

K
Finding the South Celestial Pole

Unfortunately, in the Southern Hemisphere there is no visible star within one degree of the south celestial pole. Hence, the position of the south celestial pole cannot be found with the same precision by the navigator as in the case of the north celestial pole. However, it can be found approximately.

SOUTHERN CROSS: CRUX. Finding the famous Southern Cross, a constellation called by its Latin name CRUX in the *Nautical Almanac,* is one of the best ways of locating the south celestial pole by the stars.

ACRUX: GACRUX. As shown in Fig. 13, both Acrux and Gacrux are two of the brightest stars in the constellation CRUX. Other names for these two stars are α Crucis and γ Crucis (Greek letter followed by the Latin genitive of the constellation in which the star is located). β Crucis is another bright star in this same constellation, but the fourth star is fainter, as indicated by the smaller black dot in Fig. 13. As noted in Fig. 13, a straight line from Gacrux through Acrux, if continued, will

lead almost to the south celestial pole, marked by $\oplus$ in the figure. Distances must also be noted. The distance between Acrux and the south celestial pole is about four times the distance between Acrux and Gacrux.

ACHERNAR. To help in this identification, Achernar, the brightest star in the constellation ERIDANUS, is not far from the straight line through Gacrux and Acrux, but on the *other* side of the straight line from the south celestial pole. It should also be noted that Achernar is about as far from the south celestial pole as the star Gacrux.

MAGELLANIC CLOUDS AND RELATION TO SOUTH CELESTIAL POLE. To check still further the identification of the position of the south celestial pole, the Magellanic Clouds are most valuable. Yachtsmen might prefer them. Although the Magellanic Clouds are actually neighboring systems of stars, or galaxies, they look to most people like two clouds among the brighter stars in the sky, or something like two small parts

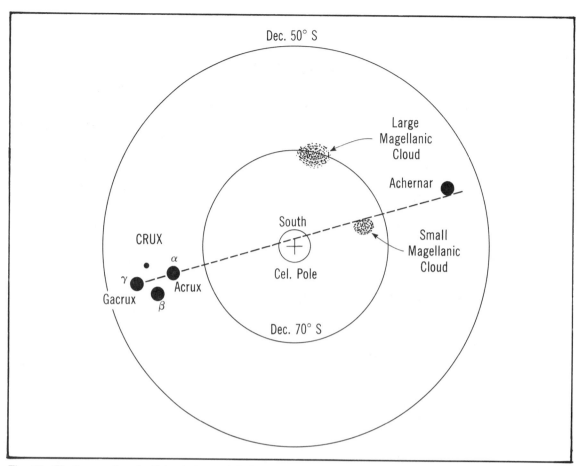

Fig. 13. Finding south celestial pole by straight line through Gacrux and Acrux, and also by using the Magellanic Clouds.

of the Milky Way, which includes most of the stars in our own system of stars. At very great distances, clusters of faint stars resemble hazy clouds. Inspection of Fig. 13 shows that if the navigator thinks of a triangle with all sides approximately equal, and places the Magellanic Clouds at two of the vertices of this triangle, then the south celestial pole will be at the third vertex, in the direction of the constellation CRUX.

The foregoing paragraph will indicate to the yachtsman that he cannot use the south celestial pole to determine his latitude to within even 60 miles, but on the other hand, this approximate knowledge would indicate the latitude to within a few degrees (200 or 300 nautical miles). Of much greater value, it will indicate the approximate direction of south in the Southern Hemisphere when the region of the north celestial pole cannot be seen at all.

PART III

Use of Sextant and Corrections to be Applied to Hs

HOLDING THE SEXTANT. A navigator uses a marine sextant in celestial navigation in order to measure the altitude of a celestial body above his horizon. As seen in Fig. 14, the navigator sees

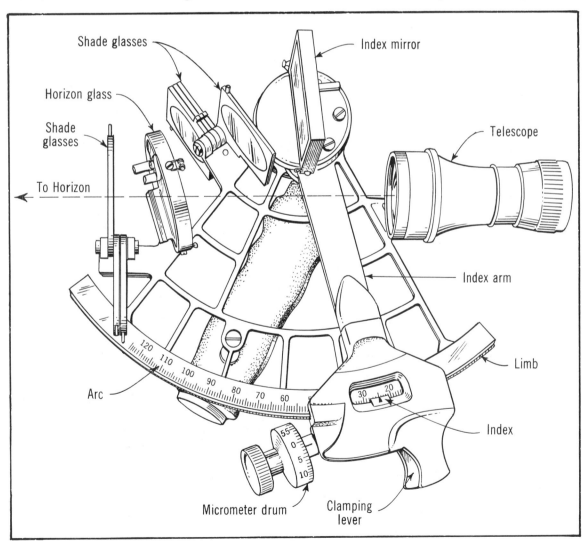

Fig. 14. The sextant.

his horizon directly through the clear glass of the horizon glass. He moves the index arm of the sextant until the light from the celestial object is reflected from the index mirror onto the horizon mirror, the silvered half of the horizon glass. Then the navigator sees simultaneously the celestial body, in the horizon mirror, and the horizon in the clear glass. On some sextants a recent development has replaced the mirror and the glass of the horizon glass with just one coated glass. While this may facilitate the taking of sextant sights, it is not a necessary change. This one coated glass does not change sextant sight procedure.

tangent, or just touching the horizon. By this "rocking" of the sextant, the navigator ensures that he is measuring the altitude on a vertical circle, and his altitude will then be accurate. The sextant altitude obtained in this manner is designated Hs. (In Fig. 15 note error if sun is not "brought down" on vertical circle by "rocking.")

Ho FROM Hs. The Hs (sextant altitude) as read from the sextant directly must be corrected to obtain the Ho (observed altitude) or true altitude. There are usually three corrections for an altitude of a celestial body, with the exception of the moon. The three corrections are shown in

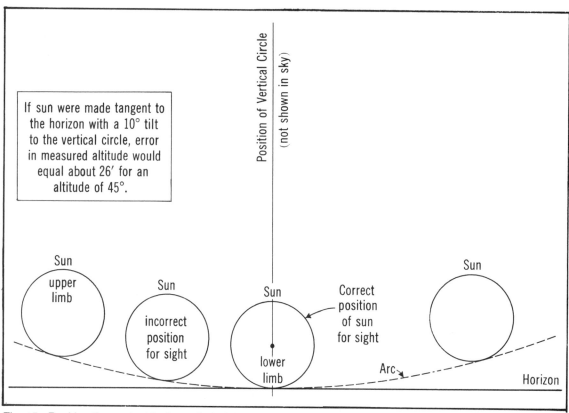

Fig. 15. Rocking the sextant for lower limb sight. (If sun, in incorrect position, is then made tangent to the horizon, the measured altitude would be too large.)

ROCKING THE SEXTANT. He must now "rock" his sextant gently. He must turn the sextant in his right hand until he sees the sun (assuming the sun is the celestial object) moving up on an arc to the left, and then up on an arc to the right, as indicated in Fig. 15. With the sun at the *lowest* point on the arc the navigator then turns the micrometer drum slightly and makes the lower limb of the sun (for a lower limb sight) exactly

Table I. They appear in the block entitled "Reduction of Sextant Observation" on the form.

Table I. Corrections to Hs

		+	−
Index Corr. (I. C.)	(1)	. . . .	. . . .
Ht. of Eye Corr. (or "Dip" Corr.)	(2)		. . . .
3rd Corr. to Hs	(3)	. . . .	. . . .
	Totals		

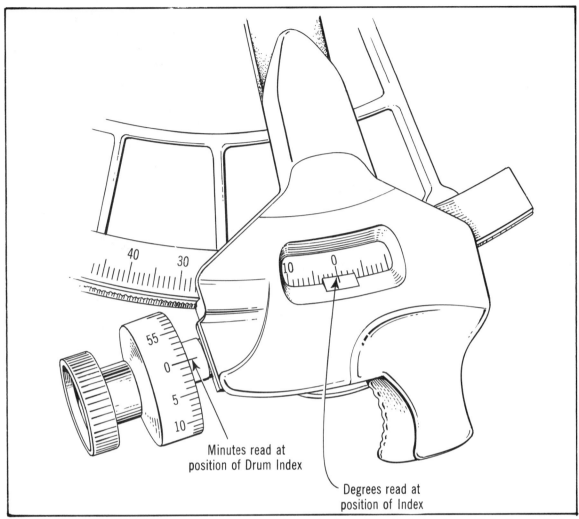

Minutes read at position of Drum Index

Degrees read at position of Index

Fig. 16. Sextant reading of 0°00.0′. (This sextant can be read only to nearest minute of arc, with tenths of minutes estimated by eye.)

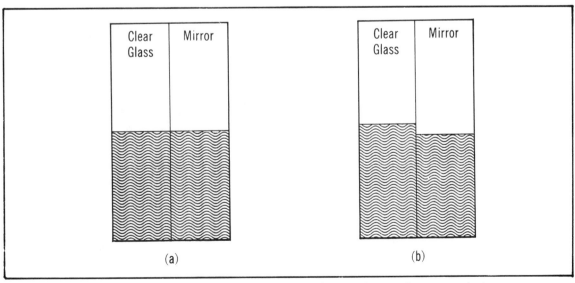

Fig. 17. Horizon glass and mirror. (Horizon glass may be round or rectangular.)

I. E. (INDEX ERROR). The first correction is due to the I. E. (index error) of the sextant. To investigate this error, adjust the sextant clamp and micrometer drum so that the reading is zero (see Fig. 16). Look at the horizon, with or without an eyepiece, but with shade glasses protecting the eye from the sun's rays. Line up the horizon, as seen in the mirror and the clear glass, so that *one* continuous line is seen, as in Fig. 17a and *not* as in Fig. 17b. Reading of sextant then is the I. E.

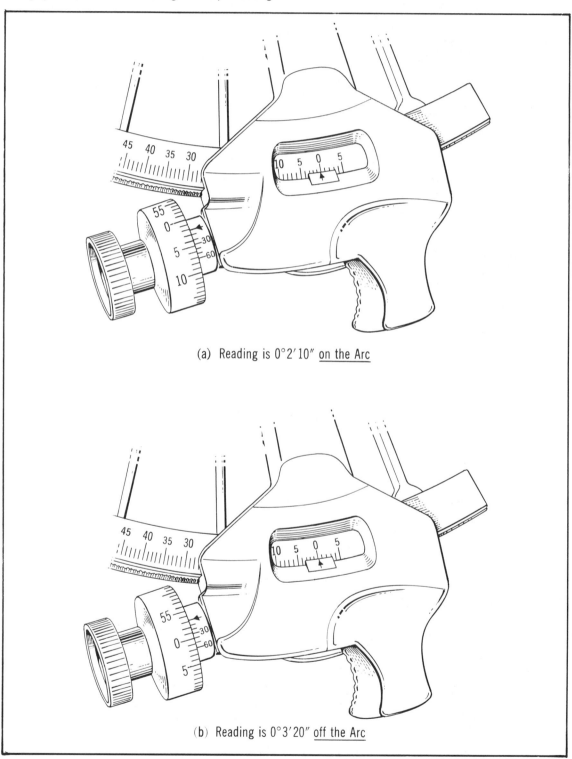

(a) Reading is 0°2′10″ <u>on the Arc</u>

(b) Reading is 0°3′20″ <u>off the Arc</u>

Fig. 18. Sextant reading which can be read to the nearest 10″ of arc.

READING THE SEXTANT. In most sextants of today the smallest division on the arc of the sextant is 1°, so that the number of degrees in any reading is taken from the sextant's arc, with numbers usually given for every 5° or 10° to aid in reading. The reading is taken at the place indicated by the arrow on the index arm. We certainly hope that the reading for any index error will be less than 1°, and so next we read the minutes and seconds. The micrometer drum has 60′ (60 minutes) of arc marked off, indicated by numbers for every 5′ of arc. Again, we take the reading in minutes at the position of the arrow on the micrometer drum. In the example shown in Fig. 18a, the reading so far would be 0°2′ and some seconds. To read the exact number of seconds, note the place on the drum where some "minute" line on the drum is an exact continuation of some "second" line to the right, on the so-called vernier. In our example here, the first line on the vernier, to the right, is exactly lined up with a minute line on the drum, to the left, and so the complete reading of the Hs in this example is 0°02′10″. Note, also, that our reading, at position of the arrow on the sextant arc, is on the main part of the sextant arc, or on the main scale. If, on the other hand, when we had lined up the horizon as seen in the clear glass and the horizon mirror, we had noted the position of the arrows as shown in Fig. 18b, the reading would have been 0°03′ 20″ *off the arc.* In the first case, with reading *on the arc,* the index correction is to be subtracted from any immediate sextant reading Hs, since this subsequent reading should be read from the new position of the sextant arm arrow, whereas we should still be reading from the numbers on the sextant arc and drum just as given. The navigator may note that when the mirrors are exactly parallel, the reading should be exactly 0° because of the design and construction of the sextant. (See Bok and Wright, *Basic Marine Navigation,* Chapter 12, for optical principle of the sextant.) Temperature changes can affect the metallic parts of the sextant, but if the navigator is careful about checking his index error to obtain his index correction before and after measuring the altitude of a celestial body, he is taking care of these small errors and will obtain an accurate Ho to within a minute of arc (equivalent to 1 nautical mile, as shown later).

In the second case of the foregoing example (reading *off* the arc), the index correction would be entered in the form as +3′20″ on the side of plus corrections in Table I, p. 22.

DIP. CORR. (DIP CORRECTION) OR HT (HEIGHT) OF EYE CORRECTION. The second correction (2) in Table I is due to the navigator's height of eye. In order to have no correction for height of eye, he would have to be taking a sight while in swimming, with his eye at water level. Above the water level he sees farther over the celestial horizon, and his Hs of a celestial body as seen above the visible or apparent horizon (Angle B in Fig. 19) is larger than it would be above his celestial horizon (Angle A in Fig. 19). The correction for height of eye is always negative. It equals 4.3′ for a height of eye of 20 feet (6 meters), and is found for other heights in the *Nautical Almanac* on the inside front cover. It is called the "Dip Corrⁿ." at the top of the page in the *Nautical Almanac.*

ALTITUDE CORRECTION. Here also, on the inside front cover of the *Nautical Almanac,* is found the 3rd Corr. to Hs (third correction to sextant altitude). It is given in the first column for the sun, and in the second column on this same page for the stars and planets.

REFRACTION CORRECTION. For the stars and planets the chief correction is for the refraction of light as it enters the denser medium of the atmosphere. Everyone has noticed how a stick looks bent when it is immersed in water; the same effect takes place with the light from a celestial body. The light is bent, or refracted, toward the perpendicular to the surface. Hence, the star, or planet, seems to be higher above the horizon than it actually is, and the navigator measures an altitude which is too large. Thus, the correction for refraction is always negative. For the star or for a planet, the refraction correction is the principal correction to the Hs, and is the "Corrⁿ." given on the front A2 page of the *Nautical Almanac* in the block headed Stars and Planets (see p. 76 in Part X). There are small additional corrections for the planets Venus and Mars, the two planets closest to the earth. They appear in the same block.

However, for the sun, the 3rd correction to the Hs, given in the first column of the A2 (Altitude Corrections Tables, p. 76), rather conveniently combines for the navigator the three corrections:

1. Refraction correction (mentioned above)

2. Semidiameter correction

3. Parallax correction

The navigator does not have to worry about these corrections separately. He just reads the total correction which corresponds to the App. Alt. (Apparent Altitude) or to the Hs obtained on his sextant. He might note that there are two columns of corrections—one for October-March Sights, and the other one for April-September

error happens to be several minutes of arc. As the *Nautical Almanac* suggests at the bottom of the A2 page (see p. 76), App. Alt. is not exactly the Hs, but it is the Hs with the first two corrections (for dip and index error) already applied to the Hs. Usually this can be done mentally. A quick glance can be given to the "dip" correction, which is usually the same on any one cruise, and also for I. E., which does not vary by more than 1' or 2' of arc as a rule. The glance will show wheth-

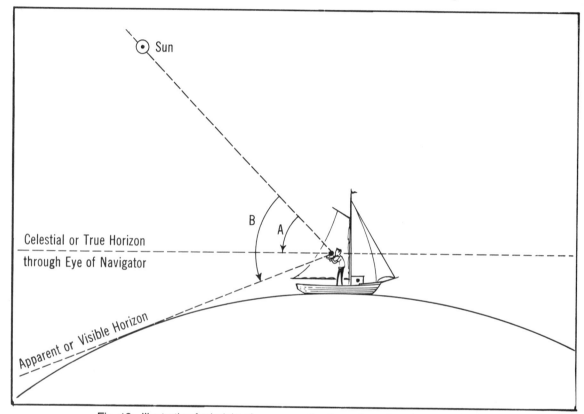

Fig. 19.   Illustration for height of eye correction (curvature of earth exaggerated).

Sights. In case the navigator enters with what is called a "critical value" for Hs (or for App. Alt.), and finds that he has two possible values for corrections, an upper and a lower one, he takes the *higher* value, in position on the page. Again this is not too serious, as the error in case of wrong choice cannot exceed 0.1'. (However, a successful navigator practices constant vigilance and strives for greatest accuracy.)

Another point is worth mentioning in case the yachtsman wants the very greatest accuracy with sextant sights. It is important if the index

er the Hs corrected for I. E. and dip throws the App. Alt. into another interval of figures. Usually, and especially if he is pressed for time (or energy), the yachtsman can forget this small item and use Hs and App. Alt. as if they were equal.

SEMIDIAMETER CORRECTION. In case the yachtsman is interested in the two other separate corrections, the most important one is the semidiameter correction, of approximately 16', equal to half the diameter or to the radius of the sun. All reduction tables and tables of position are based upon the centers of celestial bodies, as well as upon the center of the earth. The navigator cannot use the center of the sun when he takes a sight on the sun. The Hs would not be accurate if he did! Instead, he uses the lower limb (or upper limb in case of necessity). The use of limbs gives greater accuracy. Having used the limb instead of the center, however, he must correct for this substitution, and it is easy to see that the correction is about 16' (Fig. 20).

In case of (a) 16' must be added to the Hs as a correction since the sun's *center* has not been brought down to the horizon. In case of (b) 16' must be subtracted from the Hs as a correction since the sun's center has been brought down too far. *The yachtsman must note the two col-*

corrections for refraction, semidiameter and parallax.

PARALLAX CORRECTION. In general terminology parallax, the third correction, is the change in direction of an object as seen from two different positions. The object in sextant sights is the celestial body sighted. The two positions are the center of the earth (assumed in tables) and the place of observation on the earth's surface. The moon, being the earth's closest celestial neighbor, has the largest parallax of all the celestial bodies. Special tables for the moon (p. 88) are on the back cover of the *Nautical Almanac;* further explanation of the special tables for the moon are given in Part VIII. A correction for paral-

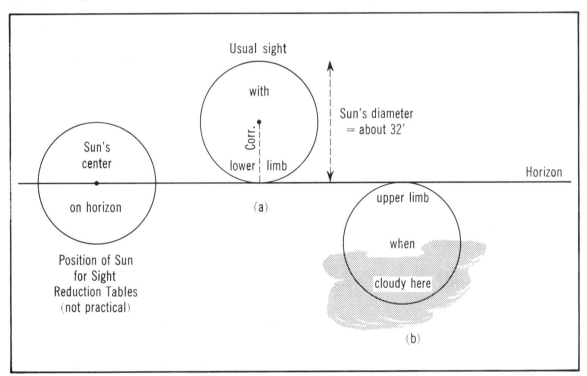

Fig. 20. Semidiameter correction with usual sight (lower limb) equals about 16', to be added to Hs.

umns in the Nautical Almanac: Lower Limb and Upper Limb (see p. 76) and enter the appropriate one. Entering the wrong column here could cause an error of 33' or 33 nautical miles. Lower limb sights are apt to be more accurate because of "rocking the sextant." For this reason they are most popular and most often used. The value of 16' changes slightly during the year because the earth's orbit is an ellipse rather than a circle with the sun at the center, but the two A2 columns for the sun take care of these minor changes, and present to the navigator the sum total of the A2

lax has to be included in the *Nautical Almanac* for the sun. The parallax correction amounts to nothing for the more distant stars, but is shown in the *Nautical Almanac* for the two closest planets, Venus and Mars. This correction and a correction for phase (we do not always see the entire *half* of Venus and Mars from the earth as in case of the moon's phases) are combined to give the rather small values found in the Stars and Planets block of the front page of the *Nautical Almanac.* The appropriate "Corrⁿ." for altitude and time of year should be included as an additional correction to those listed in Table I. (See Example 3.) The parallax correction is so important for the nearby moon that special tables for the moon are given on the back cover page of the *Nautical Almanac.* Examples in obtaining Ho from Hs follow:

1. The lower limb of the sun was sighted on May 2, 1983. The sextant reading Hs was 66° 58.8'; the index error was 8.6' *off the arc.* Find the observed altitude Ho if the navigator's height of eye was 55 feet.

Answer:

	+	–	
I. C.	8.6'		Hs = 66°58.8'
Dip Corrⁿ.		7.2'	+16.9
3rd Corr.	15.5		Ho = 67°15.7'
Totals	24.1	7.2	
Total Corr.	16.9'		

2. Sirius was sighted on February 2, 1983. The sextant reading Hs was 45°46.8'; the index error was 5.9' *off the arc.* Find the observed altitude Ho if the navigator's height of eye was 30 ft.

Answer:

	+	–	
I. C.	5.9'		Hs = 45°46.8'
Dip Corrⁿ.		5.3'	−0.3
3rd Corr.		0.9	Ho = 45°46.5'
Totals	5.9	6.2	
Total Corr.		0.3'	

3. Venus was sighted on August 20, 1983. The sextant reading Hs was 18°39.7'; the index error was 8.8' *on the arc.* Find the observed altitude Ho if the navigator's height of eye was 20 ft.

Answer:

	+	–	
I. C.		8.8'	Hs = 18°39.7'
Dip Corrⁿ.		4.3	−15.2
3rd Corr.		2.9	Ho = 18°24.5'
Add. Corr.	0.8'		
Totals	0.8	16.0'	
Total Corr.		15.2'	

(Additional Corr. is found on A2 Altitude Correction Tables, Part X, p. 76.)

4. The upper limb of the sun was sighted on a cloudy day, July 15, 1983. The sextant reading Hs was 67°25.6'; the index error was 5.4' *on the arc.* Find the observed altitude Ho if the observer's height of eye was 40.1 ft.

Answer:

	+	–	
I. C.	5.4'		Hs = 67°25.6'
Dip Corrⁿ.	6.1		−27.8
3rd Corr.	16.3		Ho = 66°57.8'
Total Corr.	27.8'		

USE OF SUN FOR SEXTANT SIGHT. In general, if the sun's limb is clearly defined, even through clouds, the sun can be used for accurate sextant sights from 1ʰ after sunrise to 1ʰ before sunset. Its altitude can be measured even near sunrise and sunset, but the refraction correction is less accurate then, and it is safer to wait for an altitude of around 15° or more, unless a lower altitude is the only possibility. When there is precipitation or complete overcast, the sun can usually be seen through breaks in the clouds or even through thin clouds if the navigator or yachtsman has the patience to stand ready, sextant in hand, for a quick sight.

Sir Ernest Shackleton's 1916 *Boat Journey* from Elephant Island to South Georgia Island in "the most tempestuous storm-swept area of water in the world" was one of the most famous and heroic boat journeys ever accomplished under extremely adverse circumstances. From the *James Caird* in the Drake Passage, the navigator Worsley obtained a necessary line of position by taking the mean or average of many sextant sights of the sun's blurred image through the clouds; there was so much fog one day that the sun had only a hazy outline. (An approximate position is better than no position at all! This particular position, however, was good.) The whole journey described in Lansing's book, *Endurance,* should be encouraging for those who think sextant sights are almost impossible in small boats (*James Caird* was only 22½ feet in length). With the sea heaving about him in the roughest ocean in the world, Worsley sometimes had to sit on the helmsman's seat and be held in place by two of the other men when he took sextant sights.

In calmer water the navigator may brace himself for a sight as shown in the illustration, p. 29.

In calmer water the navigator may brace himself for a sight as shown in the illustration. Photo: Morris Rosenfeld & Sons.

In rougher weather many navigators will agree with M. W. Richey. In his 19-foot, 9-inch *Jester* on a single-handed cruise to the Azores and back, in 1966, Richey preferred a position provided by the central circular hatch. His waist was supported there, and he had no fear of falling overboard.

PART IV

Latitude from a Noon Sight
(Meridian Passage of the Sun)

At this point a yachtsman should fully understand a noon sight, which gives a navigator his latitude at the time of meridian passage of the sun. For illustration, imagine that a navigator is at a DR

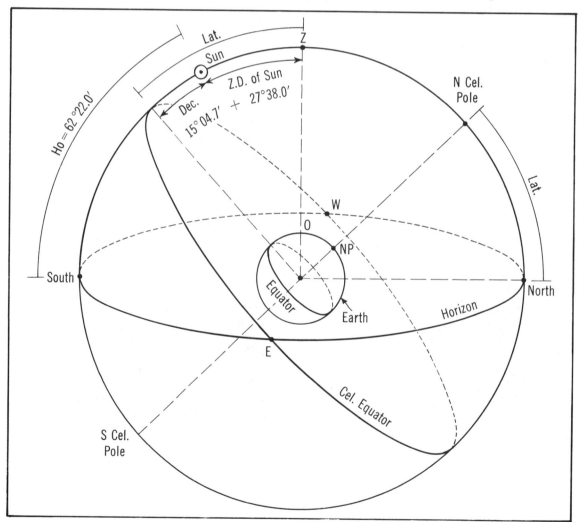

Fig. 21. Noon sight at meridian passage of sun; DR latitude, 42°44′ north when sun has declination of 15°04.7′ north and Ho of sun = 62°22.0′.

(dead reckoning) latitude of 42°44.0′ north on a day when the declination of the sun is 15°04.7′ north. If, when the sun crossed the meridian, the navigator obtained an altitude (Ho) for the sun of 62°22.0′, then the zenith distance of the sun equals 90° − 62°22.0′ or 27°38.0′. The zenith is north of the sun. If the navigator adds the zenith distance of the sun to the sun's declination at that moment, he obtains the zenith distance of the celestial equator, or 42°42.7′. According to the First Basic Relation (Part I, p. 11), the latter quantity equals the latitude of the navigator at that time (see Fig. 21). The general rule and procedure for determination of latitude from a noon sight follows.

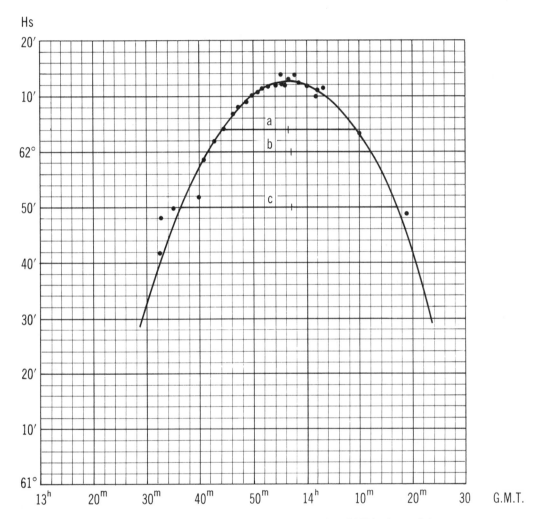

Fig. 22. Noon curve (Hs plotted vertically against GMT horizontally).

L
Method 1: "Noon Curve" Method, Valuable for all Navigators and Yachtsmen

1. For time, use watch on GMT (Greenwich mean time) or ZT (zone time) if preferred; see Part VI.

2. Measure Hs of sun; measure Hs at least every 5 minutes or more often during an interval of half an hour or more before meridian passage, when sun's altitude is highest. Continue to measure, every minute or oftener, for about 3 minutes when the sun is "hanging" on the meridian, at about the same Hs. Continue 5-minute interval (or 2-minute interval) sights for half an hour (or more) after meridian passage of sun. In case of doubt, begin taking sights at 11:00 or 11:15 zone time (Part VI), but use GMT in following plotting:

3. Plot the Hs vertically against the time, as in Fig. 22, and draw a *smooth symmetrical* curve through the plotted points, as in the same illustration. The largest value of Hs gives the altitude of the sun when it is crossing the celestial meridian, and is the value desired for determination of latitude.

4. Reduce the largest Hs on the *curve* to obtain corresponding Ho (see Form No. 2, p. 106 and Part III, p. 22).

5. Find corresponding zenith distance of sun by subtracting the Ho from 90°, and mark it N or S according to whether the zenith is north or south of the sun at that time.

6. Find the sun's declination from the *Nautical Almanac* at the time (GMT) when the sun had the Ho obtained in Step 4 above (see Form No. 2, p. 106).

7. Add zenith distance and declination if both are N (north) or if both are S (south) and obtain the latitude (N or S) as the case may be. If, on the other hand, the zenith distance was S and the declination N or vice versa, you would subtract one from the other and the answer or resulting latitude would have the same name (N or S) as the name of the larger quantity. Examples for determination of latitude from a noon sight follow:

Example No. 1: Drawing a picture or rough sketch of the celestial sphere and the position of the sun on the celestial meridian serves as well as the rule above, but not everybody enjoys drawing a picture. It is generally easier to use a rule and a form (see Part X for Form No. 2) especially after the navigator understands the theory, or possibly if the navigator of a small vessel is feeling seasick!

$$90° = 89°60.0'$$
$$\underline{Ho = 62\ 22.0}$$
ZD of sun = 27°38.0′N (zenith N of sun)

Sun's Dec. = $\underline{15\ 04.7\ N}$
Lat. = 42°42.7′N

Example No. 2: If the navigator obtained from the noon curve the largest value of the sun's altitude above the southern horizon and found Ho = 20°30.5′ at a time when the sun's declination was 18°50.2′S, his latitude follows:

$$90° = 89°60.0'$$
$$\underline{Ho = 20\ 30.5}$$
ZD of sun = 69°29.5′N (zenith observed to be N of sun)

ZD = 69°29.5′N
Sun's Dec. = $\underline{18\ 50.2\ S}$ (*Nautical Almanac*)
Lat. = 50°39.3′N

Example No. 3: If, on another day of the year, the sun's altitude was measured above the northern horizon at meridian passage and found to have Ho = 25°06.7′ when the sun's declination was 19°40.3′N, the zenith would be observed S of the sun, and the arithmetic would be as follows:

$$90° = 89°60.0'$$
$$\underline{Ho = 25\ 06.7}$$
ZD of sun = 64°53.3′S

Sun's Dec. = $\underline{19\ 40.3′N}$ (*Nautical Almanac*)
Lat. = 45°13.0′S

VALUE OF NOON CURVE. The importance and value of the noon curve of Hs for the sun for an hour around meridian passage cannot be overestimated for the junior navigator. It can also be of great value to the experienced navigator in an emergency. Consider the curve of Hs shown in Fig. 22, with Hs and GMT as indicated. The advantages are enumerated here:

1. Greater accuracy is obtained for the latitude. An accuracy to within 0.2-0.4 nautical miles or even less can easily be obtained with practice in sextant sights and use of this curve. The mean of several sights is always more accurate than just one or even two single sights.

2. A fair latitude may still be obtained if clouds gather at time of meridian passage, as often happens, by extrapolating the curve, especially if some sights after meridian passage are possible. Better yet, the mean of some of the sights already taken can be reduced by *Pub. No. 249* (Part V, Section P), and an accurate line of position before noon is possible.

3. The approximate longitude may also be obtained from this noon curve of sextant sights, as follows:

a. Obtain the axis of the Hs curve around noon (a partial ellipse). To obtain the axis, bisect several horizontal lines, such as a, b, and c in Fig. 22. This can be done experimentally by swinging of dividers, or by actual measurement. Adopt the mean of the 3 "mid"-times, as read from the horizontal axis.

b. Note the time of meridian passage from noon curve. This is the actual GMT of meridian passage at position of your vessel, if watch error is applied.

c. Use this GMT of noon to enter the *Nautical Almanac* and find the corresponding GHA of the sun at this instant. (GHA and LHA are explained in Part V, Section N; Part VII, Requirement (7), noon curve (p. 56) and Form (p. 96) show details for cruise of the *Ruchbah*.)

If longitude is west, the GHA of sun just obtained gives the longitude. In equation, longitude (west) = GHA of sun − LHA of sun, LHA of sun = 0 because the sun is *on* the meridian at this time.

If longitude is east, the GHA of sun, taken from the *Nautical Almanac,* will be found to exceed 180°, and the GHA must be subtracted from 360° to give the correct longitude east.

The accuracy of the longitude derived by this method depends for the most part upon the accuracy of the time used. The watch may have a greater error than the navigator realizes, especially if he has neither radio nor time signals, and also if he has no chronometer. In general, longitude determinations depend upon accurate time. In addition, the junior navigator might not be skillful enough in drawing the smooth symmetrical Hs curve of Fig. 22, and in finding the best axis of the curve. Practice improves accuracy.

The advantage of this Hs noon curve technique is that no special reduction table (such as *Pub. No. 249, 229, 214, 211* or *208*) is needed. Even an experienced navigator might find this reduction table missing in some emergency. (See Part V, Section S for Reduction Tables.)

The disadvantage of the noon curve technique is that it takes more time (nearly an hour) but to the navigator who enjoys taking sextant sights, the advantages are important, and the method is just one other easy way to obtain the greatest accuracy upon all occasions and also to practice constant vigilance!

M
Method 2: For Experienced Navigator

Some yachtsmen will not enjoy this computation. It can be appreciated more fully after study of Part VI.

PRECOMPUTATON OF MERIDIAN PASSAGE. For the experienced navigator, if he feels rather confident about clear skies at noon, he will precompute the time of meridian passage of the sun. At that time he can take two almost simultaneous sights of the sun and reduce the mean of these two sights to obtain an accurate latitude, as above. The details of precomputing the time of meridian passage follow:

At any time before time of meridian passage, the experienced navigator takes off a few minutes to compute the time of meridian passage of the sun. He may choose any convenient time as long as it *is* actually before the transit or meridian passage time. For example,

Step 1: At DR Lat. 42°51.0′N, Long. 29°24.0′W, on August 12, 1983, he might choose 11^{h}24^{m}00^s (GMT). (See Part VI, Section W.) He would enter this time on the Noon Sight Reduction Form (Form No. 3, p. 107). If his watch was 50^s slow, the correct GMT at this moment would then be 11^{h}24^{m}50^s.

Step 2: The LHA of the sun at 11^{h}24^{m}50^s must be found next. First the GHA of the sun can be found in the *Nautical Almanac* and entered on the form in the second block. In our example GHA of sun − Long. W. = LHA of sun = 320° 32.3′ (see Form No. 3, p. 97 for the details).

Step 3: The latter value is then subtracted from 360° to find the angular distance *t* (39°27.7′) through which the sun must travel before it is on the meridian. If the speed of the vessel is greater than 12 knots, as in this cruise of the *Ruchbah* described in Part VII, the speed of the vessel must be considered. The problem is concerned with how fast the sun is approaching the meridian. Two speeds are involved. One is the apparent speed of the sun (15°/hr.) westward, due to the earth's eastward rotation at that rate. The other is due to the vessel's speed, which may increase the first speed if the vessel is proceeding east, or decrease it if the vessel is proceeding west. If, in our example, the vessel is on a true course of 100°, speed 22 knots, the vessel's speed eastward is increasing the apparent westward speed of the sun. It is necessary to know the vessel's change, or difference in longitude per hour (D Lo/hr.), on Form No. 3 in upper right-hand corner. This D Lo/hr. can most easily be found by measuring on the chart where the course is plotted. In our example, the speed of 22 knots determines the distance of 22 nautical miles covered in an hour. On an ocean chart, or on a Universal Plotting Sheet on which the true course 100° is plotted, 22 nautical miles may be measured along the course. The change, horizontally, in longitude, or the difference in longitude (D Lo/hr.) may be measured with dividers quite simply. In our example D Lo/hr. = 30'E (Fig. 23).

and −58.1' − 18.3', or -76.4' (−1°16.4') is the exact Corr. to t.

Since our value of t is a little less than 40°, simple inspection shows that −76' is a fair correction to apply to t.

Step 5: The corrected t is, therefore, 38°11.3'. Expressed in time, the corrected t is $2^h32^m45^s$. If the interval to noon, or to meridian passage, is added to the original time $11^h24^m00^s$, the predicted watch time (GMT) of noon, or of meridian passage, is found to be $13^h56^m45^s$.

Step 6: The junior navigator should note that the predicted watch time of noon is the time

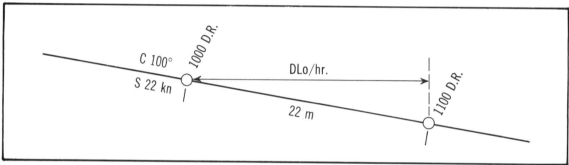

Fig. 23. Determination of D Lo/hr. (22 nautical miles measured along course shows change in longitude for the hour, D Lo/hr., which can be measured on the longitude scale of the chart used).

In the example, the arc t through which the sun must move is smaller than it would be if the vessel were anchored (or proceeding slowly), and the correction must be subtracted from t, or from 39°27.7' in the example.

Step 4: The Table of Corr. to t, given at the bottom of Form 3, p. 97, is inserted there to give the necessary correction quickly when it is needed. Some navigators will want to perform double interpolation in entering Table of Corr. to t. The average yachtsman will probably not be exceeding 12 knots and hence the Table of Corr. to t in Step 4. will be unnecessary for him. It is included for completeness. If he ever needs it, he can enter the Table comfortably, as follows, with t = 39°27.7' and D Lo/hr. = 30'E. This value of t is almost 40°. On the row of 40° for t, note that for 30'E the Corr. to t is −77.4', while for t = 30° the Corr. to t is −58.1'. The exact proportion would be:

which the watch *will read:* For the predicted GMT of noon, another small block on the Noon Sight Form is provided. Here the *correct* GMT of the original time is added to the interval to noon. The predicted GMT of noon in this case is $13^h57^m35^s$. This is the correct time to be used in obtaining the declination of sun at meridian passage. An error in seconds of time of noon does not affect the accuracy of latitude obtained unless the speed is abnormal. (The details, on forms, of this particular meridian passage on August 12, 1983, are presented in Part X on the Noon Sight Form, p. 97).

The yachtsman, or reader, may now turn to the Practice Cruise in Part VII and work out the details on the Form for Reduction of noon sight. (Part X, p. 107.) Using extracts from the *Nautical Almanac* and filling out the Forms are most important for the junior navigator before he goes to sea!

$$10 \left\{ \begin{matrix} 30° \\ 39.5 \\ 40 \end{matrix} \right[\; 9.5 \; \left. \begin{matrix} -58.1' \\ \cdots\cdots \\ -77.4 \end{matrix} \right] x \right\} 19.3'$$

$$\frac{9.5}{10} = \frac{x}{19.3} \; \text{or} \; x = -18.3'$$

PART V

Descriptive Astronomy Necessary for an Understanding of Taking and Reducing a Sextant Sight in the Morning or Afternoon

Definition of Local and Greenwich Hour Angles; Finding Hour Angles

First, it is necessary to define and understand hour angles of celestial bodies. In the Northern Hemisphere, if a navigator faces South, he can imagine his celestial meridian as it runs through his zenith and the south point of his horizon. If any celestial body was on the navigator's celestial meridian at this time, it would have an LHA (local hour angle) of 0°, since the celestial meridian (upper branch) is the starting point for measurement of local hour angles. For instance, the

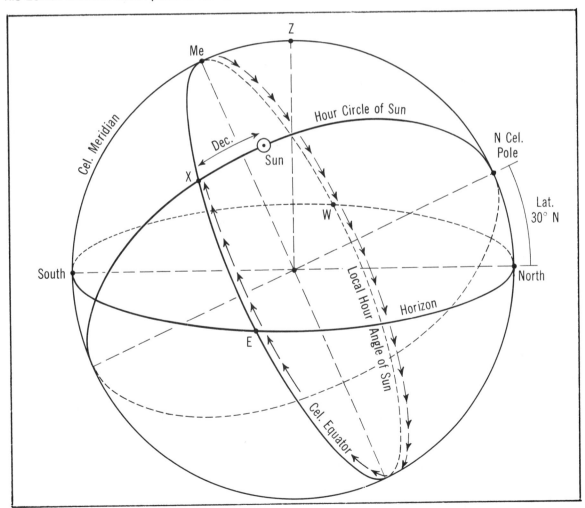

Fig. 24. Local hour angle of sun; sun east of meridian.

sun has an LHA of 0° when it crosses the navigator's celestial meridian at the "noon" meridian passage of the sun, as in Fig. 21. One hour later, the sun will have moved 15° to the west (due to the earth's real eastward rotation of 15°/hr.), and the LHA of the sun will then be 15°. An hour angle always increases to the *westward* up to 360° as the celestial body returns to the celestial meridian. The *direction* of motion as the hour angle increases is always parallel to the celestial equator.

ward along the celestial equator, way around through W and then below the horizon, up through E to the point X. At X the hour circle through the sun (similar to a meridian through any point on the earth) cuts the celestial equator. The LHA of the sun in Fig. 24 has a large value, between 270° and 360°, close to 320°. The navigator never has to measure this hour angle in practice, for which he can be duly grateful! Starting with the exact time of any sight or observation, he can easily derive the local hour angle

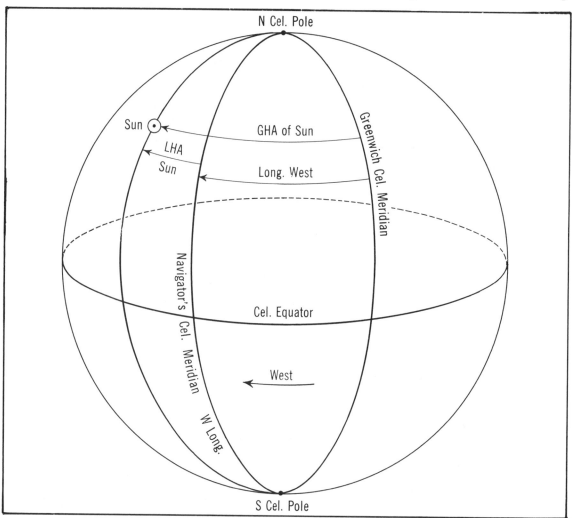

Fig. 25. a) Relation between GHA and LHA of sun. West longitude; (GHA − long. W = LHA)

For example, if at an assumed latitude of 30°N the sun was east of the celestial meridian, as shown in Fig. 24, on a day when its declination was 10°N, the navigator could roughly estimate its LHA. He could, in imagination, start at the upper branch of the celestial meridian near the celestial equator, at M_e. He could measure west-

from the GHA (Greenwich hour angle) in the *Nautical Almanac*.

The *Nautical Almanac* would be too big a volume if it gave local hour angles for every position on the earth. Hence, in order to produce a volume of convenient size, the *Almanac* makers give the GHA of celestial bodies.

The Greenwich hour angle would be the local hour angle if the navigator were at Greenwich; and from the GHA (Greenwich hour angle) it is very easy to obtain the LHA (local hour angle) of any celestial body at any place on the earth at any time. The fundamental relations to remember are:

1. For west longitude, GHA (Greenwich hour angle) of celestial body − Long. W. = LHA (local hour angle) of celestial body

2. For east longitude, GHA of celestial body + Long. E = LHA of celestial body

Relations of **1.** and **2.** can be understood by inspection of Fig. 25.

In Fig. 25a and b the horizon is not shown because hour angle relations are more clearly emphasized when the celestial equator is in a horizontal position.

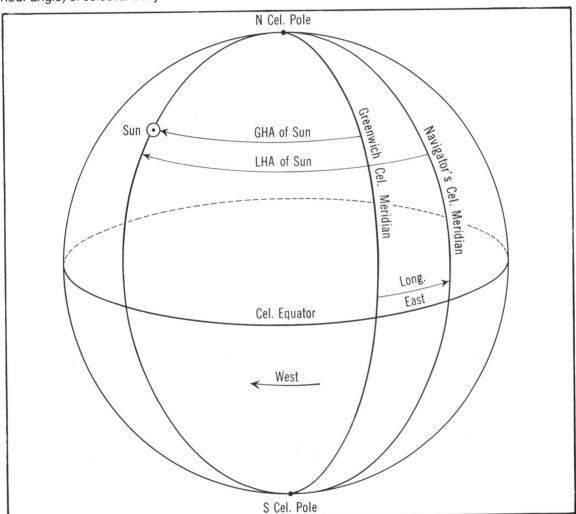

Fig. 25. b) East longitude; (GHA + long. E = LHA)

O

Explanation of Circle of Position: Its Use by Chichester

At this point the junior navigator should understand that a "line of position" is fundamentally the arc of a circle of position. The latter is sometimes called "Sumner's circle of position" because in 1837 Captain Sumner of Boston first thought of the principle involved. To understand the circle of position, it may help the junior navigator to think of a flagpole of known height H (Fig. 26). For an assumed distance (d) from the flagpole one can calculate the vertical angle Ac subtended by the flagpole. In other words, he can find the computed vertical angle between the top and the bottom of the flagpole. If at the navigator's position at O the vertical angle of the flagpole is measured with a sextant (As) corrected by index error to Ao, he can compare Ao with the computed altitude (Ac). (Note: O indicates (in Fig. 26) the point where a navigator is standing.) For example, there are three possibilities:

1. Computed
altitude (Ac) = measured altitude (Ao).
2. Computed
altitude (Ac) < measured altitude (Ao).
3. Computed
altitude (Ac) > measured altitude (Ao).

FINDING CIRCLE OF POSITION. The navigator can find his circle of position from the sun: Subtract from 90° the Ho which corresponds to the Hs of the sun's altitude at any moment. The navigator must understand that:

1. 90° − Ho, converted to minutes of arc, represents in nautical miles the radius of a circle of position derived from the sun at this instant.

2. The center of the circle of position has its latitude equal to the declination of the sun at time of sextant sight.

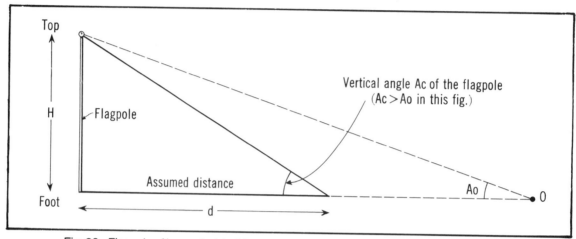

Fig. 26. Flagpole of known height (H) at assumed distance (d) from O, position of observer.

If the first possibility should occur, the navigator would know that he was at the assumed distance (d) from the flagpole. If the second possibility should hold, the navigator would be closer to the flagpole than the assumed distance (d). If the navigator found the third possibility true, as in Fig. 26, he would know that he was farther away from the flagpole than the assumed distance. In each of the three possibilities it should be noted that the navigator's distance off is not fixed at *one* point by his sextant angle. Although his sextant angle would in each case give his distance from the flagpole, of known height, the known distance off would place him somewhere on a *circle of position* with a radius equal to the distance off and a center at the foot of the flagpole.

We argue analogously for the sun; on the celestial sphere in this analogy the sun takes the place of the top of the flagpole, while the horizon replaces the foot of the flagpole in measurement of Hs. When the navigator measures the altitude of the sun above his horizon (Hs), there is a slight resemblance to measuring the vertical angle of the flagpole.

3. The circle's center has for its longitude the GHA of the sun at the instant.

RADIUS OF CIRCLE OF POSITION. To fully understand **1.** above, consider the position of the sun on the celestial sphere. In imagination project this position of the sun upon the earth. The sun on the celestial sphere falls upon the subsolar point on the earth. The zenith distance of the sun falls upon the distance on the earth between the navigator and the subsolar point. Hence, the sun's zenith distance in minutes of arc is equal to the radius (in nautical miles) of the circle of position (see Fig. 27).

In order to plot a circle of position, if he chose to do so, the navigator would need to know the latitude and longitude of the center of the circle, as stated in **2.** and **3.** above.

LATITUDE OF SUBSOLAR POINT OR OF CENTER OF CIRCLE OF POSITION. The latitude is easiest to understand. If we project the subsolar point and its distance from the earth's equator upon the celestial sphere, we have the sun and its angular distance from the celestial equator. By definition the latter angular distance is just the declination of the sun. The navigator can easily find this declination for any instant in the *Nautical Almanac*. Hence the sun's declination gives the latitude of the subsolar point (Fig. 27).

The longitude of the subsolar point, as stated in **3.** above, is understood also by projection from

can be particularly useful to the navigator. The idea was noted and used in 1931 by Sir Francis Chichester who in his youth was the first man to fly alone from New Zealand to Australia across the Tasman Sea. As the distance on this famous flight was too great for a nonstop flight because of the small fuel capacity in a small plane, Chi-

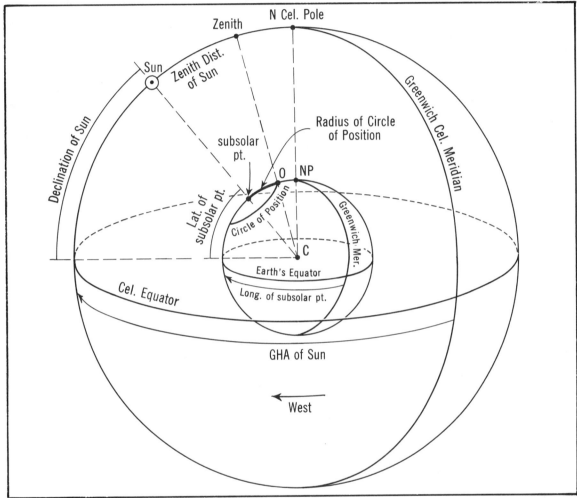

Fig. 27. Circle of position: subsolar point at the center; latitude of center = sun's declination; longitude of center = GHA of the sun.

the earth onto the celestial sphere. The longitude of the subsolar point on the earth is its angular distance from the Greenwich meridian. This distance corresponds, on the celestial sphere, to the angular distance of the sun from the Greenwich celestial meridian, or is equal to the GHA of the sun (see Fig. 27).

USE OF CIRCLE OF POSITION BY CHICHESTER. On special occasions, when headed approximately toward or away from the sun (or any other celestial body), the circle of position

chester decided to use Norfolk and Lord Howe islands as stepping-stones. He could not afford to miss these small islands, like pinpoints in the vast ocean. First there was Norfolk Island, with no radio direction-finding signals at this time. Hence, *before* departure from New Zealand, he did some computing. For the time of 1 hour before his estimated arrival (ETA) at Norfolk Island, Chichester figured out the circle of position which would pass through this island (adopting the island for his assumed position; for details of this technique, see Section P following). The radius of this circle gave him his distance (90° − Hc) in nautical miles between the subsolar point and Norfolk Island or between the subsolar point and any other point on this same circle of position. He then selected a second point 2 on this same circle of position but 1 hour of flying time from his destination and so far to one side of Norfolk Island that he would be sure which side he was on in case he was off course in proceeding to point 2. Point 2, to the west of Norfolk Island, would also be at the same distance from the subsolar point or at the distance of (90° − Hc) in nautical miles. He planned, when point 2 was reached, to change course by 90°, flying with sun abeam to port for the next hour (radius of a circle is at right angles to arc or approximate chord of the circle; see Fig. 28). The most critical time was the moment when he would be at point 2, when he would change course 1 hour away from Norfolk Island. This was the crucial and exciting moment! In order to check when point 2 was reached, Chichester planned to take sextant sights of the sun beginning 2½ hours *before* he reached point 2, when the sun was about dead ahead. If his Ho of the sun was too small to make 90° − Ho exactly equal to the computed radius (90° − Hc) of his circle of position, he would know that he still had not reached point 2. He would also know the number of nautical miles still to be

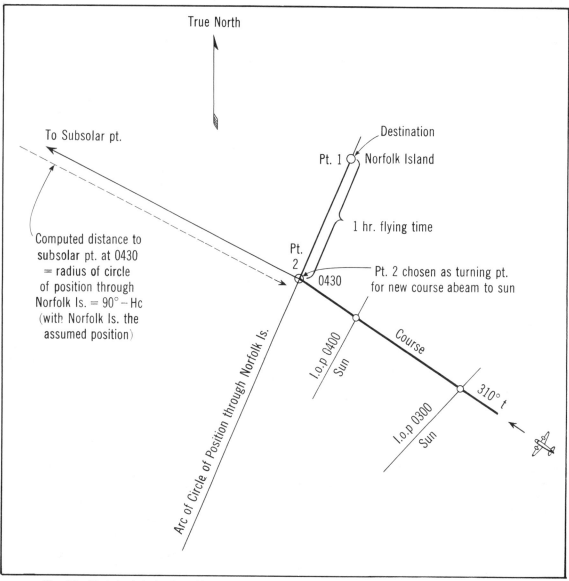

Fig. 28. Sketch of circle of position scheme used by Chichester to assure arrival at Norfolk Island.

flown to reach point 2. The number of miles would be Hc − Ho. If, later, he had found Ho larger and such that (90° − Ho) was smaller than the computed distance of the subsolar point— Warning! He would have turned back immediately. But Chichester did not allow this error to occur. He took sun sights ½ hour before the estimated time of arrival at point 2, and again at point 2 or at the estimated time of arrival there, to check that he was really there. With this technique of using the circle of position, and also by correcting his headings for the wind, Chichester successfully reached the pinpoint of Norfolk Island, his first-stop destination. Note that the actual subsolar point does not have to be plotted. Knowing the radius of the circle, or the distance of the subsolar point from Norfolk Island is sufficient. The *direction* of the subsolar point is indicated by the azimuth of the sun at the time.

It should be noted that a circle of position can be obtained by use of any of the other celestial bodies, at appropriate times, but for the moment we are concentrating on the sun, the most convenient and popular of all the celestial bodies because it can be easily seen even through some clouds or between clouds.

CIRCLE OF POSITION NECESSARY WHEN RADIUS IS SMALL. When the radius of the circle of position is small enough (sun near zenith) so that it can be plotted on a chart conveniently, it is the most accurate curve of position to use. Usually, however, the radius is too large to plot conveniently, and then the navigator will prefer to use the ingenious technique devised by a young French officer, Marcq Saint Hilaire, in 1875. Explanation of this technique follows.

P
Marcq Saint Hilaire Method

If the yachtsman looks at Fig. 29, he can easily understand the theory involved in the reduction of a morning sextant sight of the sun. In this figure, the sun is about 20° above the horizon, at S in Fig. 29, after having risen above the horizon at Sr. The navigator can measure on his sextant the Hs of the sun very accurately at this instant, and of course he can easily obtain the corresponding Ho of the sun at this time, as described in Part III.

ASSUMPTION OF MARCQ SAINT HILAIRE METHOD. A very clever trick is employed in celestial navigation—to find a line of position from this sextant sight. The navigator makes an *assumption*. He assumes that he is at some convenient position, close to his DR position at the time. He then computes, with the aid of the *Nautical Almanac* and reduction tables, what the altitude of the sun would be *if* he were at this position. The computed altitude is designated Hc. The navigator next compares his Ho with the calculated Hc. If the Ho and Hc agree in value, the navigator knows that his line of position (on which the navigator must be situated) passes through the assumed position. Usually Ho does not equal Hc (see Section Q).

Q
Explanation of Navigation Triangle for North Latitude

The reduction tables above depend upon a very important navigation triangle which the navigator understands. It is drawn in Fig. 29 for a particular time at a particular place in the Northern Hemisphere. It has the sun (S) at one vertex, the zenith at another vertex, and the north celestial pole at the third. It follows that the navigation triangle has the following sides, which are arcs of great circles:

1. ZP = co Lat. = 90° − assumed Lat.

2. SP = 90° + Dec. (S) of sun. (For north Dec. in north Lat., SP = 90° − Dec. of sun)

3. SZ = 90° − Hc = 90° − computed altitude of sun. Sides (1) and (2) are known as soon as the latitude is assumed. The angle *t* between sides (1) and (2) can be found for any given time. With the above three quantities known, the navigation triangle can be solved, and side (3) determined. In this way the altitude which the sun would have if the navigator was at the assumed position can be found.

In detail, for an assumed latitude, the navigator knows the value of (1). Next, for any given time, the declination of the sun may be found in the *Nautical Almanac*. Hence, side (2) is obtained. The spherical angle *t* has the same measure as its arc QQ_m in Fig. 29, and the arc QQ_m equals 360° minus the larger arc westward from Q_m (360° − LHA of sun at the instant). We have seen above that the GHA of the sun for any instant is given in the *Nautical Almanac*. From the GHA, the LHA of the sun is easily computed by subtracting a W longitude or by adding an E longitude. The longitude in this case is the *assumed* longitude. The navigator now knows two sides and the included angle *t*. Hence, the navi-

gation triangle as pictured in Fig. 29 can be solved and the third side SZ determined. If SZ is known, it is easy to obtain Hc, the altitude which the sun would have if the navigator was at the latitude and longitude assumed above.

FOOT-POINT. Solution of the navigation triangle also gives the angle at Z *in* the triangle. When the celestial body is east of the meridian at north latitude, as in Fig. 29, this angle is also the azimuth (denoted by Zn in *Pub. No. 249*) of the sun. In Fig. 29 the azimuth of the sun is the arc NF, or the angle of Z in the navigation triangle. The point F can be called the "foot-point" of the sun on the horizon.

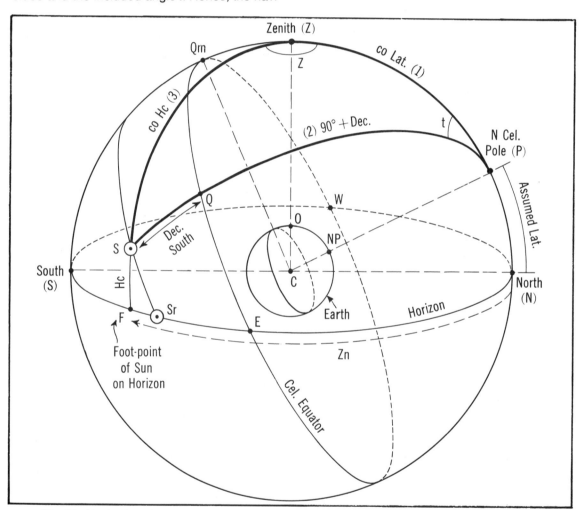

Fig. 29. Navigation triangle for sun in morning at north latitude. (Azimuth of sun = Zn = Z in triangle.)

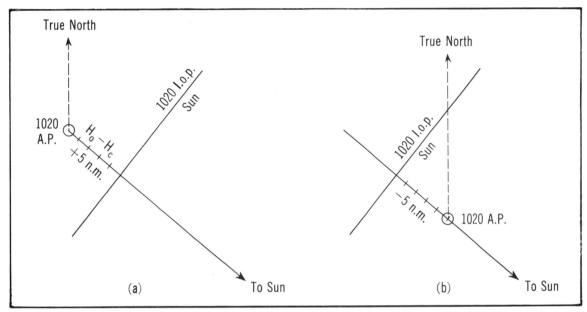

Fig. 30. (a) Example of LOP for an azimuth of 130° and intercept of +5 nautical miles at 1020; (b) Example of LOP for an azimuth of 130° and intercept of −5 nautical miles at 1020.

R
Plotting of Line of Position and Explanation

To plot the LOP (line of position) explained above:

1. Plot the assumed position (A. P.) with time noted beside the point.

2. From this A. P. plot the azimuth of the sun (Zn) at the time of the sun sight.

3. Find Ho − Hc, the *intercept*.

If Ho > Hc, Ho − Hc is positive, and the navigator is closer to the subsolar point than his A. P. (assumed position).

If Ho < Hc, Ho − Hc is negative, and the navigator is further away from the subsolar point than his A. P.

If Ho = Hc, the line of position goes through the assumed position.

In each of the three cases, the LOP is drawn at right angles to the azimuth line, and it is drawn at the end of the intercept (Ho − Hc), which has been measured off in nautical miles (Fig. 30a and b).

The above directions for plotting an LOP are stated on the Forms. The navigator realizes, of course, that the LOP is the arc of a circle of position; if the circle is very large, the arc is practically a straight line, as in Fig. 28.

In the afternoon, when the sun is west of the celestial meridian, the angle at Z in the navigation triangle gives the azimuth only indirectly. Fig. 31 for the same north latitude as in Fig. 29 shows the sun at S in the afternoon of the same day, and west of the celestial meridian. The angle Z in the navigation triangle has the same angular measure as the short arc NF in the figure. Hence, the angle at Z, found by solution of the navigation triangle, must be subtracted from 360° to give the true azimuth (denoted by Zn in *Pub. 249*) of the sun according to the rule given on the Form. The angle *t* equals, directly, the LHA of the sun at the time.

S
Sketches of Navigation Triangle for South Latitude

Figures 32 and 33 are shown to clarify the rules on the *Pub. 249* Form for south latitude. The South Pole must be drawn *above* the south point of the horizon with altitude equal to assumed latitude. The basic Rule I (see p. 11) still holds, and the arc ZQ_m (where Z represents the zenith) must also equal the assumed latitude. If the declination of the sun happens to be south, as in Fig.

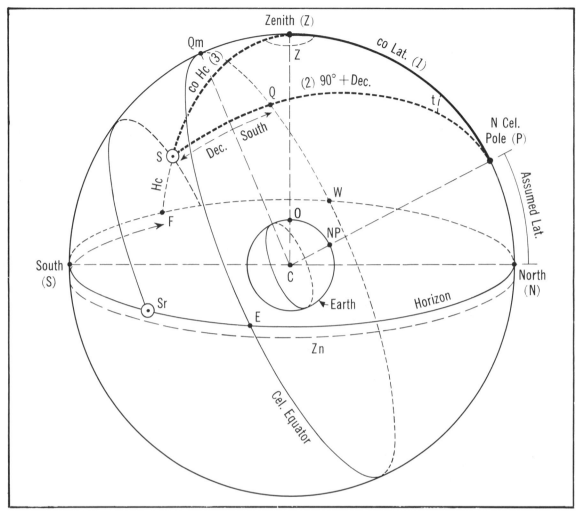

Fig. 31. Navigation triangle for sun in afternoon at north latitude. (Azimuth of sun = Zn = 360° − Z in triangle.)

32, for a morning sun sight with the sun east of the celestial meridian, the navigation triangle still exists. The zenith is at one vertex, the south celestial pole at another vertex, and the sun itself at the third. The angle Z in this triangle equals the arc SF on the horizon, and the arc SF on the horizon equals 180° minus the arc NF, or 180° minus the true azimuth of the sun. Hence, when the sun is east of the celestial meridian, the true azimuth of the sun equals 180° minus the angle Z in the navigation triangle. For LHA of the sun, if the yachtsman wanted to trace it out in his imagination, he would start at Q_m and travel along the celestial equator *westward*, or through W and E back to Q. As in the Northern Hemisphere, for a morning sun sight, the LHA of sun is between 180° and 360°, but now the azimuth Zn (NF in Fig. 32) equals 180° minus angle Z in the triangle. For an afternoon sight, as in Fig. 33, LHA of sun is

between 0° and 180° in value, and the true azimuth Zn equals 180° + Z.

Every navigator should realize that many techniques have been developed for a solution of this important triangle. An almost complete list is given in Bowditch's *American Practical Navigator (Pub. No. 9)*. Each technique has special advantages and disadvantages; a few are now specified.

SOLUTION OF NAVIGATION TRIANGLE BY TRIGONOMETRY. Firstly, the navigator who is familiar with spherical trigonometry should remember that the law of sines and the law of cosines along with a table of trigonometric functions (or a calculator) are adequate for the solution of the navigation triangle. No other reduction tables are absolutely necessary. (See author's *Particularized Navigation* in References, p. 119.) Other fast reduction tables actually do most of the reduction work for the navigator, who merely reads the results after he finds the appropriate page and column in which to read.

SOLUTION OF NAVIGATION TRIANGLE BY H. O. PUB. NO. 211; BY H. O. PUB. NO. 208. Secondly, Ageton's Method (*H. O. Pub. No. 211*) and Dreisonstok's Method (*H. O. Pub. No. 208*) have the advantage of presentation in small volumes of a few pages. They have the disadvantage of requiring more reduction-time (by a factor of 3 or 4). They are fully explained in Bok and Wright's *Basic Marine Navigation*, which also

emergency tables and are included (as Table 35) in volume two of *Pub. No. 9, American Practical Navigator,* Bowditch. (A condensed or modified Ageton Method, *Compact Sight Reduction Table,* by Allen Bayless is published by Cornell Maritime Press.)

SOLUTION OF NAVIGATION TRIANGLE BY H. O. PUB. NO. 214. Thirdly, for many years the *Tables of Computed Altitude and Azimuth (H. O. Pub. No. 214)* have been both fast and accurate reduction tables, and widely used. A disadvan-

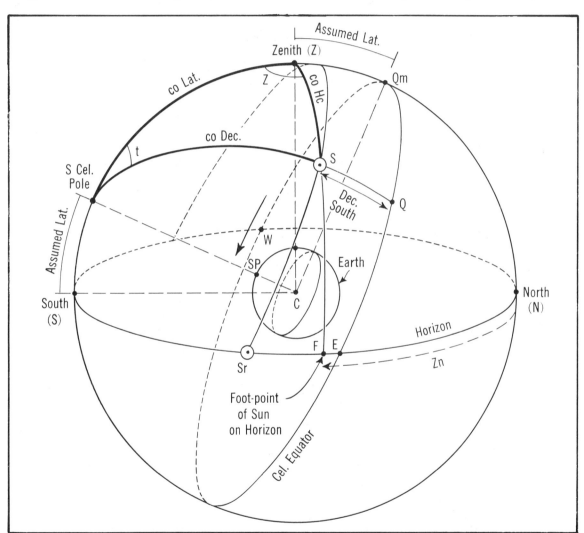

Fig. 32. Navigation triangle for sun in morning at south latitude. (Azimuth of sun = Zn = 180° − Z in triangle.)

contains the complete tables of *H. O. Pub. No. 211*. These tables are strictly mathematical; they are still accurate and never outdated, though they are no longer published as a separate volume by the U. S. government. They are good

tage arises because nine volumes are necessary for world coverage. Each volume includes data for ten degrees of latitude, and there is limitation to altitudes greater than 5°. *Pub. No. 214* has now been phased out.

SOLUTION OF NAVIGATION TRIANGLE BY PUB. NO. 229. In order to offer an improvement over *H. O. Pub. No. 214,* the Defense Mapping Agency Hydrographic Center (DMAHC) publishes newer tables, *Pub. No. 229.* "The aim has been to achieve absolute accuracy with a high standard of design and presentation," according to Blythe, Duncombe, and Sadler.

SOLUTION OF NAVIGATION TRIANGLE BY PUB. NO. 249. Finally, *Sight Reduction Tables for Air Navigation (Pub. No. 249),* in three volumes, are the tables chosen for sight reductions

from use of *Pub. No. 249,* as compared with *H. O. Pub. No. 214* or the newer *Pub. No. 229.* Most navigators are willing to accept the loss of tenths of a mile if they gain time and save energy in sight reductions, and yachtsmen with other duties may be more tempted to take more sights if the reductions are quick and easy. Volume 1, in particular, designed for stars, offers the fastest reduction of any tables to date. Although originally intended for air navigation, *Pub. No. 249* has become very popular with marine navigators, and is widely used today. Volume 2, for

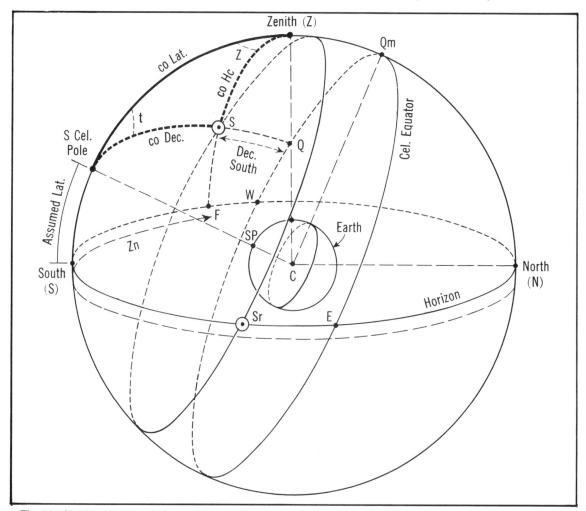

Fig. 33. Navigation triangle for sun in afternoon at south latitude. (Azimuth of sun = Zn = 180° + Z in triangle.)

in this instruction manual, for several reasons: First, they have sufficient accuracy for most yachtsmen. Tenths of minutes are not given in the values for Greenwich hour angles, nor in the Table of Correction to Tabulated Altitude for Minutes of Declination which helps with interpolation. Hence, an error of about a mile might arise

latitudes 0°-40° and declinations 0°-29°, is used for sight reductions (at latitudes indicated) of the sun, moon, planets, or any stars not in Volume 1 but with declinations within the 0°-29° interval. Volume 3, for latitudes 39°-89° and declinations 0°-29°, was planned for sight reductions at higher latitudes, as indicated, and for similar celestial bodies. It should be emphasized that *Pub. No. 249* can be used with the *Nautical Almanac* just as conveniently as with the *Air Almanac,* which also can be used by the marine navigator. The *Air Almanac* has the disadvantage of being printed in two volumes a year; the marine navigator often is at sea for many weeks, and the change from one *Air Almanac* to another and the orders and details of purchase involved cause most marine navigators to prefer the *Nautical Almanac,* printed yearly.

In summary, the chief publications for celestial navigation follow. *Pub. No. 249* is published by DMAHC and may be obtained from the DMA Office of Distribution Services or its authorized sales agents. The British version of *Pub. No. 249 (A. P. 3270)* is published in a case binding. This binding may satisfy the needs of those yachtsmen who find the comb binding unsuitable in the wet environment of an open boat.

For stars we need *Pub. No. 249,* vol. 1; for sun, moon, planets, and extra stars, *Pub. No. 249,* vol. 2 (latitudes 0°-40°), and *Pub. No. 249,* vol. 3 (latitudes 39°-89°). *The Nautical Almanac* (for current year) is purchasable from the Superintendent of Documents, U. S. Government, Washington, D. C. 20402.

H. O. Pub. No. 208 and *H. O. Pub. No. 211,* published as a separate volume, were phased out during 1970-72. Former *H. O. Pub. 211* is now Table 35 in vol. 2 of *Pub. No. 9* (Bowditch). *Pub. No. 229* provides the most accurate sight reduction tables, but *Pub. No. 249* may be easier to use and is more valuable with stars.

PART VI

Explanation of Different Kinds of Solar Time

An explanation is now given, in the same order, of:

1. **Apparent time,** either local (LAT) or Greenwich (GAT)

2. **Mean time,** either local (LMT) or Greenwich (GMT). GMT is also called **Universal Time** (UT)

3. **Universal Time Coordinated** (UTC)

4. **Standard** or **Zone Time** (ST or ZT)

5. **Summer** or **Daylight Saving Time** (DT or DST)

T
Apparent Time

Apparent time depends upon the rotation of the earth with respect to the real sun. The apparent day begins when the real sun is on the lower branch of the navigator's celestial meridian. After the earth rotates eastward during 24 hours, and the real sun is on the lower branch of the celestial meridian again, one whole apparent day of 24 hours has elapsed at the position of the navigator. Although each day consists of 24 hours, with *noon always at 12ʰ00ᵐ in apparent time,* these days are unequal in length. There are two reasons for this inequality. One chief reason is due to the fact that in January the earth moves faster in its annual eastward journey around the sun, while in June the earth moves more slowly. It was Tycho Brahe's observations of planets, around 1580, which enabled the astronomer Kepler to deduce several laws about planetary orbits and motions (see Hodge's *Concepts of Contemporary Astronomy*). To understand the facts pertinent to unequal days, it is helpful to look at a schematic figure of the earth revolving around the sun. If the earth's path around the sun were a circle instead of an ellipse, the subject of time would be more simple. The elliptical orbit of the earth is much exaggerated in Fig. 34, in order to emphasize the relations. The earth is closer to the sun in January than in June. According to the law of equal areas (area I being equal to area II) the earth moves from position 1 to position 2 in the *same* time in which the earth moves from position 3 to position 4. Since it is obvious that the arc between positions 1 and 2 is shorter than the arc between positions 3 and 4, the earth must move more rapidly in January than in June. Hence, the days, defined by successive appearances of the real sun on the lower branch of the meridian as the earth rotates, are unequal. The earth has to rotate through a slightly larger angle around January between successive appearances of the real sun on the meridian, and the days are therefore slightly longer at this time.

U
LMT (Local Mean Time) and GMT (Greenwich Mean Time)

Since it would be inconvenient on the earth to have days of unequal length, astronomers invented a fictitious sun which revolves eastward uniformly, or at the same rate along the ecliptic,

48

so that the days in one year are practically equal in length. Our clocks and watches in everyday life depend upon the rotation of the earth with respect to this fictitious or mean sun. The mean day begins when the mean, fictitious sun is on the lower branch of the celestial meridian at any place on the earth. After the earth rotates eastward during the whole day, and this mean sun is again on the lower branch of the celestial meridian, one whole mean day has elapsed. The time, at any place on the earth, which depends upon these mean days, with each mean day divided into 24 mean hours, is called LMT (local mean time). When the particular place on the earth is at Greenwich, England, or on the Greenwich meridian, the local mean time there is designated as GMT (Greenwich mean time). The difference between the GMT and the LMT is the longitude of the place. Add the longitude west, or subtract the longitude east, to or from the LMT of the place to obtain the corresponding GMT.

EQUATION OF TIME. The difference between any specific local mean time and the correspond-

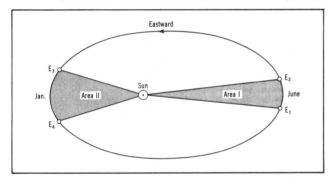

Fig. 34. Earth's elliptical orbit (exaggerated) to illustrate Kepler's Law of Equal Areas.

ing local apparent time is defined as the equation of time. Fortunately for the navigator, the equation of time never exceeds about 16½ minutes. Hence, the yachtsman with a small vessel and possible steering error of 5° can forget about the equation of time when he uses *Pub. No. 260* for checking his compass by the azimuth of the sun. He can enter *Pub. No. 260* with the LMT as if it were the local apparent time, and he finds the LMT from the GMT which he keeps on a watch or chronometer, as explained below. A navigator may navigate without ever using the equation of time. However, some navigators use it occasionally, as with *Pub. No. 260*. (See Part IX.) The equation of time is given on every daily page of the *Nautical Almanac*, for every 12 hours of GMT.

Table II

Day	Sun		
	Eqn. of Time		Mer. Pass.
	00ʰ	12ʰ	
11	05ᵐ19ˢ	05ᵐ14ˢ	12ʰ05ᵐ
12	05 10	05 05	12 05
13	05 00	04 54	12 05

Interpolation may be used for greatest accuracy. Inspection of the adjoining column, entitled "Sun, Mer. Pass.," shows the LMT of the sun's meridian passage. If this time is later than 12ʰ00ᵐ00ˢ (LAT of meridian passage), for that day the LMT is always later than any corresponding LAT, and the equation of time would always be subtracted from any LMT to obtain the corresponding LAT.

Universal Time Coordinated (UTC)

Due to the unpredictable variation in the rotation of the earth, distinction must be made between two types of universal time. The time transmitted on time signals and kept on chronometers and watches is Universal Time Coordinated (UTC). The time based on the rotation of the earth, including the variations in the rate of rotation, is UT1. This is the time that corresponds to the hour angles from the surface of the earth, and is the time that is the basis of the tabulation in the *Nautical Almanac*.

UTC is kept within ±9/10 of a second by the introduction of leap seconds as necessary. The predicted difference between UT1 and UTC is transmitted by code or voice on the time signals as DUT1 = UT1 − UTC.

Unfortunately GMT and UT are sometimes used for both of these times. In common navigational use GMT means UT1. In applications of civil time, GMT means UTC. The details of the code and the resulting longitude correction are given on page 118.

V
ST (Standard Time) or ZT (Zone Time); Summer Time or Daylight Saving Time

People in everyday life do not carry GMT on their watches. They carry a time called *standard time* on land, and *zone time* at sea. This zone time is always a whole number of hours from Greenwich mean time. Just for convenience, the whole earth has been arbitrarily divided into strips called time zones, each zone being 15° (or one hour) in width. The meridian which runs through (or nearly through) the center of each strip or zone is called the central or standard meridian for that zone, and zone time for any zone is really the local mean time of the central or standard meridian. The time zone at Greenwich extends 7½° or ½ hour either side of the Greenwich meridian, and all the people in this 15° strip keep their watches on the Greenwich mean time which goes with the meridian through Greenwich. In other words, GMT is the ZT for this whole 15° strip. (For standard time exceptions, see pp. 262-65 of the *Nautical Almanac*.)

At any instant the sun can be on the lower branch of only *one* celestial meridian. If zone time had not been invented, the days on different meridians in one zone would begin at different times—most difficult for keeping an appointment! That is why it is most convenient for all the people in the Greenwich time strip to keep on their watches the GMT of the central meridian through Greenwich. We must investigate now what happens outside the time zone through Greenwich. Suppose, for instance, that the mean sun is on the lower branch of the celestial meridian of Greenwich at 00^h00^m on a certain day. During the next hour the earth would rotate eastward through 15°. One hour later, GMT, the sun would be on the lower branch of the celestial meridian which is 15° to the west of Greenwich. However, according to the definition of the beginning of any day, the day is now just begining at this meridian 1^h west of Greenwich. Hence the zone time in the time zone 1^h west of Greenwich is 1^h earlier than Greenwich time (zone time at Greenwich). In other words, 01^h00^m GMT at Greenwich corresponds to 00^h00^m in the time zone 15° to the west of Greenwich. In general, time zones over the earth keep a zone time which

is an integral or whole number of hours away from Greenwich time. The easiest method for finding the integral number of hours is to transform the longitude of the local place into hours and minutes (see Arc to Time Conversion Table, first yellow page of the *Nautical Almanac*). Then, by inspection, the nearest whole number of hours is evident. This number is defined as Z. D. (Zone Description).

For example, if the longitude of the place is 62°05′W, the equivalent in hours, by consulting the Arc to Time Table of the *Nautical Almanac,* is

$$
\begin{aligned}
62° \quad &= 4^h08^m \\
05′ \quad &= \underline{0^m20^s} \\
62°05′ &= 4\ 08 \quad \text{(disregarding the} \\
&\qquad\qquad\quad \text{seconds),} \\
&= 4 \text{ hours to the nearest whole hours} \\
&= |\text{Z. D.}|
\end{aligned}
$$

If the longitude is west, this nearest whole number of hours is subtracted from the GMT to obtain the corresponding ZT at the place. In the example above, if the GMT at longitude 62°05′W is $21^h47^m48^s$ on June 14, the corresponding ZT at the place is 4 hours less, or $17^h47^m48^s$ on the same day, June 14. If, on the other hand, the longitude is east, the number of whole hours is added to the GMT to obtain the corresponding zone time.

For another example, in east longitude, assume that at 39°58.8′ east longitude, the GMT is $23^h56^m28^s$ on June 30. Expressed in time, the longitude is found as shown:

$$
\begin{aligned}
39° &= 2^h36^m00^s \\
58.8′ &= \underline{0\ 03\ 55} \\
& \quad\ \ 2^h40^m
\end{aligned}
$$

in hours and minutes, and $= 3^h$ to nearest whole hours, and $3^h = |\text{Z. D.}|$. We must add 3^h to GMT $23^h56^m28^s$ and obtain $26^h56^m28^s$ on June 30, or $02^h56^m28^s$ on July 1. The cautious navigator will always write down the date along with any time! A complete log is important. Lack of a written statement probably caused the disaster in *Admirals in Collision,* an account by Richard Hough of what actually happened in 1893 on one peaceful day during practice maneuvers of Britain's Navy. Lack of a visual radar plot on the *Andrea Doria* was probably largely responsible for the collision of this ship with the *Stockholm* in 1956 (Alvin Moscow's *Collision Course*).

actuality. It just helps people to arise earlier in the morning.

"**HOUSEBOAT OF TIME**." To summarize the different kinds of solar time, the schematic "Houseboat of Time" in Fig. 35 may be useful in

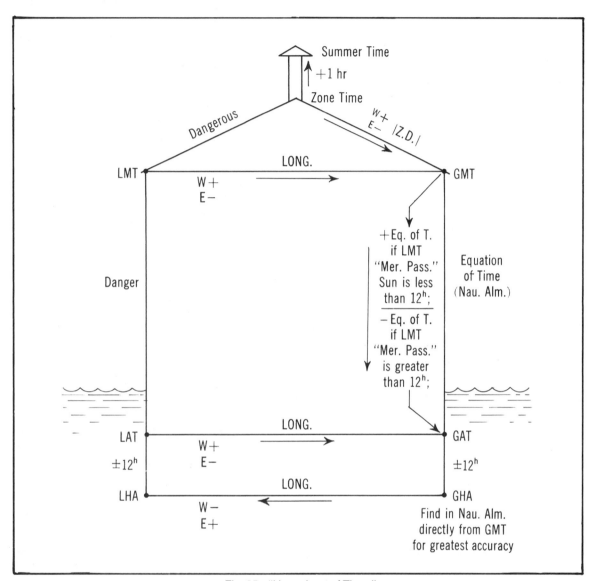

Fig. 35. "Houseboat of Time."

The last kind of time listed at the beginning of Part VI is found by adding one whole hour to zone or standard time. It was adopted so that people could enjoy more hours of daylight in the summer; hence the name "summer time," or sometimes "daylight saving time." Of course summer time does not make more hours of sunshine in

converting one kind of time to another. For directions in using the Houseboat, always go from one to another corner of the boat *around* the boundaries, as indicated by the arrows, and *never through* or across the Houseboat. Then it is easy to obtain another particular time which corresponds to any given time.

W
Setting Watch on GMT

The junior yachtsman can use the Houseboat of Time to set GMT on his watch at the beginning of a cruise. For example, if on May 1 the yachtsman took point of departure off Cape Cod at longitude 70°03'W, the correct eastern standard time might be $03^h16^m52^s$. The longitude 70°03' equals $4^h40^m12^s$ in time, or 5^h to the nearest hour. The corresponding

$$
\begin{aligned}
GMT &= 03^h16^m52^s \text{ May 1} \\
&+ 5 \\
\hline
&= 08\ 16\ 52 \text{ May 1}
\end{aligned}
$$

If the yachtsman keeps this GMT on his watch for the rest of the cruise, and just remembers about the rate of the watch, or the amount of gain or loss in one day, he may find it easier for his navigation reductions. This is certainly safer; it is the GMT which is always used in entering the *Nautical Almanac*, and a mistake in arithmetic is less likely if GMT is continually used in the reduction of celestial sextant sights. Some navigators even use GMT in plotting the course and keeping track of the DR. Personal preference is always present in many situations, but the fact remains that whatever time system is used, the navigator and the yachtsman must understand and be able to convert one kind of solar time to another. This applies especially to GMT and to LMT, and also to transferring from GMT to ZT at sea, or to standard time on land. The yachtsman will set the accurate GMT on his most accurate watch or on his chronometer, and he will treat this timepiece with the greatest of care for the whole cruise. Most timepieces keep better time if left in some protected place on a vessel, with temperature as constant as possible, and a less accurate watch may be used for the actual taking of sights on deck, and may be compared with the more accurate watch, or with the chronometer or chronometers as soon as possible. According to Tabarly's *Lonely Victory*, Hasler took four chronometers in the 1964 Atlantic Race. To quote, "I am taking four chronometers whose rates I have been checking for the past month ashore." It is even safer to take a radio as a check on the time of the most accurate watch or chronometer, and three or four timepieces are safer than one. In other words, if the navigator or yachtsman does not have an accurate timepiece, usually he cannot possibly obtain an accurate longitude by celestial navigation. Determination of longitude at sea was impossible before 1735, when John Harrison produced the first marine timekeeper. During the following nineteen years he made two more, and then continued to persevere until he constructed his celebrated No. 4 Watch, the "most famous timekeeper which ever has been or ever will be made" (R. T. Gould's *John Harrison and His Timekeepers*). It should be noted today, however, that even without a radio or time signals as a time-check, an experienced navigator may check a watch or chronometer, to within a few seconds, by accurate sextant sights of the sun (or other celestial body) *at a known position*. Worsley did this on Elephant Island before embarking on Shackleton's incredible voyage to South Georgia Island in the 22½-foot *James Caird.* Naturally, time-signals or a radio are to be preferred!

CHECKING THE TIME AT A KNOWN POSITION. In case of any emergency the navigator can obtain a line of position in the usual way with the best time available to him *at a known position*. To obtain an approximate time, he may use a variation of the noon curve method. The LMT of meridian passage of the sun, given on each daily page of the *Nautical Almanac* (see Table II) gives the navigator *at a known position* a fair time, to within 3 or 4 minutes, as soon as he expresses this LMT in GMT or ZT and then sights the sun with his sextant to obtain its highest Hs for the day (at meridian passage). This time cannot be exact because of the sun's "hanging" on the meridian for around three minutes. Next, with the best time available to him, the navigator can determine the line of position from a celestial body (usually the sun) at some other time than meridian passage. If the time is accurate, the line of position will run through the known position. If it fails to do so, the navigator can see what adjustment of time will change the Hc, and consequently the intercept Ho − Hc, so that the line of position will run through the known position. The work is really the reverse of the usual procedure.

The Ho − Hc is determined first, so as to run through the known position. Since Ho was measured, it cannot be changed; Hc has to be changed to give the required Ho − Hc. Since Hc depends upon the time of the sight, the appropriate value of Hc then determines the time. Inspection of *Pub. No. 249* shows the LHA which will give the desired Hc of the body; applying the known longitude to this LHA indicates the GHA of the celestial body used. With GHA of the body known, the corresponding GMT can be found. Details of the "inspection" mentioned will be shown in Part VIII (p. 61) in discussion of Requirements (1) and (2) of the Practice Cruise; the student yachtsman will find it easier to go through a reduction in the normal fashion first, before going into "reverse" in the procedure. It should be emphasized also that this technique works only when the position is known and the sextant sight accurate.

TIME-SIGNALS. The wise navigator will carry a radio for time-signals. Radio station WWV, Fort Collins, Colorado, broadcasts on 2.5, 5, 10, 15, and 20 MHz. Station WWVH, Hawaii, broadcasts on 2.5, 5, 10, and 15 MHz. There is a voice announcement usually near the end of every minute. The voice explains that the time is coordinated universal time, which the navigator will use as his most accurate GMT for his sextant sights. The National Bureau of Standards, Fort Collins, Colorado, welcomes reports on reception and methods of use. Station CHU is readily available in Ottawa, Canada, broadcasting on 3330, 7335, and 14670 kHz. Further details for radio time-signals are found in *Radio Navigational Aids*, in *Pub. No. 117* (RAPUB117A for the Atlantic and Mediterranean area and in RAPUB117B for the Pacific and Indian oceans area). The navigator should practice receiving time-signals *before* he departs on a cruise.

SUNRISE, SUNSET, AND MERIDIAN PASSAGE TIME. The yachtsman should note that the times given in the *Nautical Almanac* for sunrise, sunset, meridian passage of the sun, as well as time of meridian passage for all planets, are given in LMT as stated in the "Explanation" of the *Nautical Almanac*. Reference to Fig. 35 will show that it is safest to go from LMT to the corresponding GMT by applying the longitude expressed in time. If the ZT is desired, the Z. D. can be applied to the GMT to give the ZT, according to rules of the Houseboat of Time (Fig. 35).

To illustrate the concepts of time, the following example is given, with answers: Extracts from the *Nautical Almanac* may be consulted, in Part X.

At Lat. 45°10.0′S, Long. 129°56.6′E, the ZT is $06^h52^m00^s$ on August 12, 1983.

(a) Find corresponding GMT
Answer:

$$\text{Long.} = 8^h39^m46^s \text{ (arc to time)}$$
$$06^h52^m00^s \text{ Aug. 12} = 30^h52^m00^s \text{ Aug. 11}$$
$$- \quad 9$$
$$\text{GMT} = 21^h52^m00^s \text{ Aug. 11}$$

(b) Find corresponding LMT at this place
Answer:

$$\text{GMT} = 21^h52^m00^s \text{ Aug. 11}$$
$$+ \quad 8 \ 39 \ 46 \text{ Long. E}$$
$$29 \ 91 \ 46 \text{ Aug. 11}$$
$$\text{LMT} = 6^h31^m46^s \text{ Aug. 12}$$

(c) Find GHA of sun at the above time
Answer:

$$\text{For GMT} \quad 21^h52^m00^s \text{ Aug. 11}$$
$$\text{GHA of sun} = 133°42.4′ \text{ for } 21^h$$
$$+ \quad 13 \ 00.0 \text{ for } 52^m$$
$$= 146°42.4′$$

(d) Find LHA of sun at the above time and place
Answer:

$$\text{GHA of sun} = 146°42.4′$$
$$+ \text{Long.} \quad 129 \ 56.6 \text{ E}$$
$$\text{LHA of sun} = 276°39.0′$$

(e) Find Mer. Pass. (meridian passage) of sun to nearest minute only
Answer: 12^h05^m (See *Nautical Almanac*) "Mer. Pass."

(f) Find GMT of Mer. Pass. at Lat. 45°10.0′S, Long. 129°56.6′E
(At Mer. Pass.)
Answer:

$$12^h05^m = \text{LMT of sun, Aug. 12}$$
$$- \ 8 \ 39 \ 46 \quad \text{Long. E}$$
$$03^h25^m14^s = \text{GMT Mer. Pass. Aug. 12}$$

(g) Find ZT of sunrise at the same place on Aug. 12, 1983

Answer:

07^h03^m = LMT from daily page *Nautical Almanac* Aug. 12

+24

31 03 = LMT Aug. 11

−Long. 8 39 46 E

22 23 14 GMT Aug. 11

+ 9 Z. D.

31 23 14 ZT Aug. 11

−24

$7^h23^m14^s$ ZT Aug. 12

(h) Find Eqn. of T. (Equation of time) for Aug. 12, $06^h52^m00^s$ ZT

Answer:

Corresponding GMT is $21^h52^m00^s$ (above), Aug. 11. Interpolation between 05^m14^s for 12^h on Aug. 11, GMT and 05 10 for 00^h on Aug. 12, GMT gives 05^m11^s for GMT 21^h52^m Aug. 11

(i) Find corresponding GAT at the same place and time

Answer:

GMT = $21^h52^m00^s$ Aug. 11

− 05 11 Eqn. of T.

GAT = $21^h46^m49^s$ Aug. 11

Note that the Eqn. of T. was subtracted from the GMT because the LMT of Mer. Pass. is later than 12^h00^m.

(j) Find ZT of civil twilight in morning

Answer:

+ 06^h32^m Aug. 12 LMT; is taken directly from *Nautical Almanac*

+ 24

= 30 32 00 Aug. 11 LMT

− 8 39 46 Long. E

= 21 52 14 GMT Aug. 11

+ 9 Z. D.

30 52 14 ZT Aug. 11

or $6^h52^m14^s$ ZT Aug. 12

CROSSING DATE LINE. A problem in time can arise whenever the *date line* (approximately 180° from the Greenwich meridian) is crossed. Crossing this line westward causes the loss of one day; crossing it eastward gains a day according to the local calendar. To coordinate with other local people around this longitude, the navigator who cruises westward skips a day (jumping from August 11 to 13, for instance) while the navigator who cruises eastward runs through the same day twice. The theory is easy to understand if the junior navigator imagines he is in an airplane flying with the sun on his meridian. At the appropriate speed the plane can keep up with the apparent daily motion (15°/hr.) of the sun all day long. With the sun on the upper meridian for a whole day, it would be noon during the whole day. No time would have elapsed, according to our definition of time, and a day would be lost as compared with the calendar of the man who stayed at home on that same meridian. The junior navigator who flew east on a similar airplane with similar speed, from this same meridian, would have the sun on his upper meridian three times; at the beginning, during, and at the end of the complete circuit of the earth. He would have gained a day in his personal local calendar. The same results hold if many days are spent in the circumnavigation of the earth.

It is safer for the junior navigator if he keeps GMT on his watch, as stated previously. According to GMT no day is lost, nor gained, when the navigator crosses the date line. For the navigator who prefers to navigate with ZT, it is safe if he always converts from ZT to the corresponding GMT and back again to ZT, following the rules of the Houseboat of Time. With this procedure there is nothing especially new about crossing the date line. The navigator can forget about the skipping or losing of a day. He knows that the date does not change unless (as usual at any longitude) he must add or subtract 24 hours to any specific time. If he adds 24^h to any time (to avoid a negative time in subtracting the longitude, for an example) the new time is then counted from the preceding day. If he subtracts 24^h from any specific time, he is running over into the next day, and the date is changed to this next day. A combination of GMT and common sense is the best general rule for the navigator when crossing the date line, or in any other situation.

PART VII

Practice Cruise for a Day at Sea on the S. S. *Ruchbah*, with DR Checked by Celestial Navigation

Practice Cruise Requirements

The reader or navigator may work out the reductions on the blank Forms on pages 105-112. He may then look at the filled-in Forms which appear on pages 95-102, either for a check or for assistance in his own work. Necessary Extracts from the *Nautical Almanac* and *Pub. No. 249* are given in Part X for use in the reductions by *Pub. No. 249*. Two Universal Plotting Sheets are also provided on pages 113, 114, as well as graph paper for plotting a noon curve, p. 115. Answers to the requirements of this cruise are listed at the end of the cruise. If explanations are necessary, there are supplementary remarks on this practice cruise in Part VIII, in addition to the general theory of celestial navigation in the first six Parts. (Answers are at end of this Part; details on Forms are in Part X, pp. 95-102.)

You are navigator on the bridge of the S. S. *Ruchbah* out on the North Atlantic, with the following information:

Date: August 12, 1983
True Course: 100°
Speed: 22 knots, assumed for the day
Draft: 30 ft.
Height of Eye: 77 ft.
I. E. of Sextant given for each sight
Watch on GMT (watch error given for each sextant sight)

As navigator you would have examined a *Pilot Chart of the North Atlantic Ocean* (DMA Pilot 16) for the appropriate month. It would have guided you in setting courses for the cruise. It would have shown recommended lane routes, the prevailing winds, ocean currents, and marine climate for the month.

METHOD FOR OBTAINING A MORNING SUN SIGHT. At DR position: Lat.   42°53′N
                         Long. 29°40′W
you take a sun sight, lower limb, as follows:
watch (GMT): $10^h51^m17^s$; watch error: 50^s slow
Hs = 40°30.9′
I. E. = 2.5′ Off the arc

Requirement 1: Intercept and Azimuth (Zn) after filling out Form for Reduction by *Pub. No. 249*, vol. 2 and vol. 3 in Part X, p. 105.

Requirement 2: Plotting of line of position (LOP) from the sun on a Universal Plotting Sheet. No. 1 (Assume, on plotting sheet, Mid. (mean) Lat. of 43°N and Mid. Long. of 29°W.) Plot also true course 100°.

TIME OF MERIDIAN PASSAGE (NOON): A METHOD FOR EXPERIENCED NAVIGATOR. At $11^h24^m00^s$ GMT (watch time) (watch error 50^s slow) you decide to compute the GMT of meridian passage of the sun. Use the Form for Reduction of Noon Sight. Instead, if you prefer, you can soon begin taking sextant sights for the noon curve.

Requirement 3: Predicted watch time and predicted GMT of meridian passage (noon) Form No. 3, p. 97; or wait for the noon curve.

Assume that for the mean of three sextant sights of sun's lower limb, taken over a 2-minute interval, at meridian passage:
Hs = 62°12.5′
I. E. = 0.5′ Off the arc
Use predicted watch time found in Requirement 3
Watch error still 50^s slow

Requirement 4: Latitude at meridian passage, or find it from noon curve.

METHOD FOR FINDING LATITUDE WITH NOON CURVE. Begin to take sun sights, lower limb if possible, about half an hour before meridian passage. (About 11:00 or 11:15 ZT are safe times for beginning sights, in general.) Imagine that you take sun sights, lower limb, as follows:

Watch Time (GMT)	Hs	Watch Time (GMT)	Hs
$13^h32^m40^s$	61°42.2′	$13^h52^m30^s$	62°11.7′
13 33 45	61 48.2	13 53 20	62 11.8
13 35 40	61 50.2	13 55 00	62 14.2
13 40 00	61 52.0	13 55 45	62 12.3
13 41 03	61 58.7	13 56 30	62 13.3
13 43 00	62 02.4	13 57 50	62 14.1
13 44 30	62 04.6	13 58 40	62 12.6
13 46 50	62 07.4	13 59 40	62 11.6
13 47 45	62 08.7	14 00 20	62 11.8
13 48 45	62 09.3	14 02 20	62 11.5
13 49 50	62 10.2	14 03 10	62 11.8
13 50 30	62 11.0	14 10 00	62 03.4
13 51 55	62 11.3	14 19 35	61 49.0

If you are alarmed at this array of sextant sights, you may want to omit every other one. They are all shown so that the student navigator will realize that 0.2 nautical miles may be obtained for maximum error just by taking many sights. Usually the number of sights depends entirely upon the accuracy desired and also upon the skill of the navigator. The mean of several figures is always the safest! Four sights should be a minimum number if they are possible. Method in Requirements 3 and 4 is preferred by many navigators, especially by those who have other duties to perform. The noon curve method gives greater accuracy and practices constant vigilance.

Requirement 5: Plotting of the above observations against the corresponding watch times. A curve is drawn through the points merely by eye. It should be smooth and symmetrical. The sun sets just as it rises, without any bumps and loops of any kind.

Assume that the Hs read off from this noon curve is 62°12.5′; I. E. = 0.5′ Off the arc.

Requirement 6: Latitude from the noon curve. (Note: The same latitude should be obtained as in Requirement 4.)

Requirement 7: Longitude from the noon curve, as follows: Assume that three midpoints are read or measured as follows:

$13^h56^m30^s$
13 57 00 (Inspect noon curve in Part IV, p. 31
13 57 20 for comparison with yours.)
The mean of these three values is
$13^h56^m57^s$ watch time
<u>   50</u> slow = watch error
13 57 47 GMT August 12
when sun is at meridian passage.

With this GMT find from *Nautical Almanac* the GHA of sun. GHA of sun at meridian passage equals longitude W. (See p. 96, Form No. 2, for assistance.)

Requirement 8: Plotting of noon latitude from the sun, with appropriate labels, on Universal Plotting Sheet No. 1, pp. 103, 113.

Requirement 9: Running fix from sun at 1358 GMT by using the latitude line just obtained and advancing the 1052 LOP along the course.

State both latitude and longitude.

Continue plotting true course of 100° from this 1358 running fix.

OBTAINING AN AFTERNOON SUN SIGHT. In the afternoon you plan to obtain a fix from the sun and the moon. Your first step is to take three sun sights as follows:

Watch Time (GMT)	Hs (lower limb)	
$16^h18^m00^s$	48°08.7′	
16 19 30	47 59.3	
16 20 25	47 53.0	Watch error
		51^s slow; I. E.
		0.5′ On the arc

Since these sights lie nearly on a straight line when you plot the Hs against time, you decide to reduce only the last one.

Requirement 10: Intercept and azimuth (Zn for sun) after you have filled out Form for Reduction by *Pub. No. 249,* Part X, p. 108. Compare results with Form No. 4, p. 98.

OBTAINING AN AFTERNOON MOON SIGHT. Your second big step in obtaining your fix from the sun and from the moon is to take three moon sights, *upper limb,* as follows:

Watch Time (GMT)	Hs (upper limb)	
16^{h}21^{m}55^s	44°13.6′	
16 23 00	44 22.3	
16 23 50	44 30.1	Watch error = 51^s slow; I. E. = 1.5′ On the arc

Since these sights lie nearly on a straight line when you plot the Hs against the time, you decide to reduce just the first one, since it is nearer in time to the last sun sight.

Requirement 11: Intercept and azimuth (Zn) of the moon sight above, after filling out Form for Reduction by *Pub. No. 249,* p. 109. Compare results with Form No. 5, p. 99.

Requirement 12: Plotting of the two lines of position just obtained for the sun and moon, respectively, on a Universal Plotting Sheet, No. 2. See p. 104 for comparison. (Use 42°N for Mid. Lat. and 26°W for Mid. Long.)

Requirement 13: Lat. and Long. of the 1621 fix from the sun and moon.

OBTAINING TWILIGHT 3-STAR FIX. At 1621 GMT the true course 100° is plotted from the 1621 fix, speed still 22 kn. At evening twilight you obtain a 3-star fix from the following data:

Star	Watch time GMT	Hs	
Arcturus	21^{h}12^{m}50^s	48°45.3′	
Antares	21 16 08	21 07.0	
Altair	21 17 25	40 39.0	Watch error = 53^s slow; I. E. = 2.5′ Off arc

Requirement 14: Reduction of the 3-star fix on a Star Fix (*Pub. No. 249*) Form, p. 110. See p. 100 (Form No. 6) for comparison.

Requirement 15: Intercept and azimuth of Arcturus.

Requirement 16: Intercept and azimuth of Antares.

Requirement 17: Intercept and azimuth of Altair.

Requirement 18: Plotting, on the same Universal Plotting Sheet, No. 2, of the three lines of position, with appropriate labels. (Note that the Antares LOP is practically parallel to the course; in this case no advancing or retiring of Antares LOP is necessary.) For precise navigation you may want to retire the Altair LOP for the time difference between Arcturus and Altair sights.

Requirement 19: The 2114 star fix: latitude and longitude, for 1980.

Requirement 20: The Precession and Nutation Correction Table to fix for 1980. (Found in Part X, p. 94.) (See Part VIII, pp. 65, 66, for further explanation.)

Requirement 21: The 2114 star fix: latitude and longitude, for 1983. The navigator will consider that his position lies somewhere within or on the boundary of the triangle formed by the three lines of position from the three stars. He might draw a circle to enclose the triangle and represent his fix by this circle, with possible error of 1-3 nautical miles. The *central point*, by eye (either of triangle or circle), is usually adopted as a probable position of fix. It should be noted that the correction for precession and nutation just about counteracts or compensates for the retiring of the Altair LOP. Hence, the first triangle drawn (without retiring the Altair LOP in this case) would have served for a good fix and would have saved the navigator a welcome respite from extra work.

The happy and possibly weary navigator has now taken enough sextant sights for one day.

However, in order to present an example of the reduction of a planet sight, the planet Saturn might have been sighted as follows:

Watch (GMT): 21^{h}12^{m}57^s; watch error = 53^s slow
Hs = 23°34.3′
I. E. = 2.5′ Off the arc

Requirement 22: Intercept and azimuth after filling out Form for Reduction by *Pub. No. 249,* vol. 2 and 3, p. 112. (See filled-in Form No. 8, Part X, p. 102 for comparison.)

The navigator would now be wise to catch a good sleep while he can. No further accurate sights are possible until morning twilight, when he may obtain another 3-star fix if weather permits. The Practice Cruise on the S. S. *Ruchbah* ends, in this book, with the evening twilight fix obtained above at 2114 GMT.

Answers to Requirements in Cruise of S. S. *Ruchbah*

1. Ho − Hc = +7 nautical miles; Zn = 109°
2. See Plotting Sheet No. 1, p. 103
3. $13^h56^m45^s$; $13^h57^m35^s$
4. 42°42.2′N
5. See Graph of Noon Curve, p. 31
6. 42°42.2′N
7. 28°11′W
8. See Plotting Sheet No. 1, pp. 103, 113
9. 42°42.2′N; 28°07′W
10. Ho − Hc = −18 nautical miles; Zn = 241°; from A.P. 27°03.3′W, 42°N
11. Ho − Hc = −30 nautical miles; Zn = 158°; from A.P. 26°55.0′W, 42°N
12. See Plotting Sheet No. 2, p. 104
13. 42°31′N; 27°00′W
14. See Form, p. 100
15. Ho − Hc = +14 nautical miles; Zn = 250°
16. Ho − Hc = −15 nautical miles; Zn = 188°
17. Ho − Hc = −14 nautical miles; Zn = 120°
18. See Plotting Sheet No. 2, p. 104
19. 42°11′N, 24°35′W
20. Fix moved 1 nautical mile in direction 100°
21. 42°11′N, 24°34′W
22. Ho − Hc = +5 nautical miles; Zn = 232°; from A.P. 24°15.4′W, 42°N

PART VIII

Supplementary Remarks on Some of the Details of the Cruise of the S. S. *Ruchbah*

Plotting Preliminaries

PLOTTING OF TRUE COURSES AND MAGNET-IC COURSES. According to conventional usage the navigator usually plots on his chart true courses (from the true geographical North Pole). Some marine navigators prefer to use the magnetic compass rose when it is shown on charts, as on coastwise charts, but DMAHC charts of oceans cover large regions, with changes in variation. A constant value for variation across an ocean does not often exist. (See Part IX if further explanation is needed on magnetic directions.) It is safer to plot true directions in celestial navigation; on the Universal Plotting Sheets used for the August Cruise on the Atlantic, true courses and true directions are plotted. If the student navigator desires an ocean chart for this cruise, *DMAHC chart WOAGN121* is recommended. Some navigators would use the WOXZP series of Position Plotting Sheets (974 for the Practice Cruise, see p. 120). It is conventional to plot lines of position (with labels) on extra plotting sheets, in order to keep the ocean chart clean for the desired courses. A fix, denoted by a triangle (used by the U. S. Air Force and the author), or by a circle, can be transferred from a Universal Plotting Sheet to the ocean chart, such as WOAGN121. A running fix, obtained as in piloting, when one LOP has to be advanced along the course (to be used with a second LOP) is labeled R. FIX, with the time.

ENTERING GMT ON FORMS. On the Forms the block concerned with the time of the sextant observation or sight is arranged for two kinds of procedure which yield the correct GMT:

1. With GMT on the watch (as suggested in Part VI), the first two lines of the Form can be ignored. The GMT watch time ($10^h51^m17^s$) for the morning sun sight of the cruise will then be entered on the line indicated, as shown on Form No. 1, p. 95. The watch error of 50^s slow was found in this case by using the watch rate and also by comparison with a chronometer with known error, checked by radio.

2. The navigator who prefers to carry ZT (Part VI) on his watch will enter the ZT of the sextant sight on the first line of the Form. To find the corresponding GMT, he applies the Zone Description (Part VI) to the zone watch time. The Z. D. has a + sign in longitude west, and a − sign in longitude east. If for this morning sight the navigator had read $8^h51^m17^s$ on his watch, he would have filled in the Form as follows:

Zone watch time $08^h51^m17^s$
Z. D.$+$ 2
GMT watch time   10 51 17
GMT watch error   50 slow
GMT $10^h52^m07^s$
Greenwich date   Aug. 12, 1983

WATCH ERROR. The watch error of 50^s for the first sextant sight is rather large, but it is better not to change the hands of a good watch or chronometer except on rare occasions. A watch error might even involve minutes of time.

Data From *Nautical Almanac* For Sun Sight Reduction

For Requirement 1, data are taken from the *Nautical Almanac.* In filling out the Form for this sun sight, turn first to the daily page of the *Nautical Almanac* which covers August 12, 1983. The correct GMT of the sun sight is $10^h52^m07^s$. Steps of reduction procedure are as follows:

OBTAINING GHA OF SUN

Step 1: Use the daily page of the *Nautical Almanac* (p. 78) which gives GHA and declination for *hours* of GMT. Hence, in our example, the GHA of sun for 10^h is 328°43.7′ (entered on Form I, p. 95).

Step 2: Turn to yellow page of the *Nautical Almanac* (p. 86) to take care of additions due to the minutes and seconds of GMT. 52^m will be found at the top of a yellow page. Seconds of time run vertically. The addition of GHA of sun for 52^m07^s is 13°01.8′, as shown on Form 1, p. 95.

Step 3: For the sun there is no *v* value, nor *v Corr.*ⁿ

Step 4: The lines of the Form "For Stars" exist only for the reduction of stars not listed among the seven possible stars in *Pub. No. 249* for any LHA ♈. Skip over these lines in reducing a sun sight, as well as for moon and planets.

OBTAINING LHA OF SUN
FOR ENTERING *PUB. NO. 249*

Step 5: By addition of the two above GHA values, the GHA of sun for GMT $10^h52^m07^s$ is 341°45.5′, entered on Form. We need to apply the DR longitude to this GHA of sun in order to obtain the LHA of sun at the GMT of sight. In *Pub. No. 249*, however, we want to enter the tables with a whole number of degrees for the LHA (indicated on Form by 00.0′). Hence, in W longitude the minutes of the assumed longitude should equal the minutes of the GHA. (In E longitudes the minutes of the assumed longitude should be 60′ − minutes of GHA of sun.) The desired LHA of sun in our present case is 312°.

DECLINATION OF SUN FROM *NAUTICAL ALMANAC*, FOR ENTERING *PUB. NO. 249*

Step 6: When the daily page of the *Nautical Almanac* was consulted to obtain the GHA of sun, the declination of the sun for the hours of GMT should also have been written on the Form in the block "To find declination": 15°05.2′N for our present reduction. In general, whenever the daily page of the *Nautical Almanac* is consulted for one value, all other values needed later should also be written on the Form, to save the time wasted in turning back to the daily page a second time. This rule applies to any of the celestial bodies. At the same time, for the sun, the value of *d* is needed and can be written on the Form. It is found at the bottom of the daily page, in the Declination column for the sun. For our sun sight this value of *d* is 0.7. It helps us interpolate for the minutes and seconds of GMT, but before turning to the "yellow page" we need to *inspect* the declination values of the sun to see whether they are increasing or decreasing around 10^h GMT. We note in this case that they are decreasing, and hence we place a minus sign on the d Corr.ⁿ line to indicate subtraction. We find the value to subtract by consulting the yellow page for the 52^m of GMT. We find 0.7 in the (*v* or *d*) column and read 0.6 in the corresponding Corr.ⁿ column. The minus sign has already indicated subtraction (which we also know if we review our picture of the apparent diurnal motions of the sun on the celestial sphere during a year). The sun's declination is 15°04.6′N for our 1052 GMT sun sight.

ENTERING *PUB. NO. 249* SIGHT REDUCTION TABLES, VOL. 3

Step 7: Since we now have the necessary data for entering the *Sight Reduction Tables Pub. No. 249*, vol. 3 (latitudes 39° − 89°) we turn first to the page marked Latitude 43°, whole degree nearest to DR Latitude. There are six pages with Lat. 43° at the top. We want the page with Declination *Same* Name as Latitude since the declination of the sun is north and the latitude is also north. When one of these names is north, and the other south, the Declination *Contrary* Name to Latitude pages are used. In addition, at Lat. 43° there are pages for declinations from 0° − 14°, and other pages for declinations 15° − 29°. We want a page which covers 15°, p. 23 of *Pub. No. 249*, vol. 3. We are ready to enter with declination first. As we found in **Step 6** above, the sun's declination for GMT of sight is 15°04.6′ and hence this value is between 15° and 16°.

Necessity for Entering Pub. No. 249, vols. 2 and 3 with Smaller Declination: In *Pub. No. 249*, vols. 2 or 3, *always* enter the declination column with a whole number of degrees *less* than the exact declination of the sun at the time. If we had had a declination of 15°47′ we still should have entered the 15° declination column.

Enter with LHA of Body: In the 15° declination column, run down the page until the line with LHA of 312° is reached (LHA 312° was found in **Step 5** above). On this line we almost have our answer. We read here 40°29′ for the tabulated Hc, a value of *d* equal to +42, and Z equal to 109°.

CORRECTION TO TABULATED Hc

Step 8: We now have to apply a correction to the tabulated Hc because our sun's declination is not exactly 15° but 15°04.6′, with a declination difference of 04.6′ from 15°00.0′. The + sign in front of 42 signifies that we have to *add* a correction. This correction is given in the last Table in *Pub. No. 249, Vols. 2 and 3, with heading Correction to Tabulated Altitude for Minutes of Declination (see p. 92). We enter the latter Table with the d* value of 42′ at the top of the page. We run down the page vertically until we come to 4.6 (declination difference) in the right or left margin, and read 4 for our correction, which must be added, as noted above. Note that Hc is given to the nearest whole minute of arc in *Pub. No. 249*; hence Ho is finally rounded off, on Form, to nearest whole minute.

FINDING Zn FROM Z.

Step 9: To find Zn, the true azimuth to be plotted, read Direction No. 3 of Form I, p. 95. (Part V, Section Q explains the theory for the Rules.) In the present case, since the LHA of sun lies between 180° and 360°, Zn equals Z or 109°.

Step 10: The intercept Ho − Hc, necessary for plotting the line of position, is just a simple case of subtraction at this point, and equals +7′ or +7 nautical miles. A minute of arc in altitude can be expressed as a nautical mile. A nautical mile, in navigation, can be considered, usually, as the length of one minute of latitude or as one minute of any great circle of the earth, such as a vertical circle, along which altitude is measured.

Step 11: The two values Ho − Hc and Zn are necessary for plotting the line of position from the $10^h52^m07^s$ sun sight, according to the Directions

for Plotting at the bottom of the Form which are as follows:

a. Plot assumed Lat. and Long. (43°N and 29°45.5′W)

b. Lay down bearing of observed body (Zn or 109°). Draw arrow for identification.

c. From assumed position plot intercept, Ho − Hc, toward or away from bearing of body, according to whether Ho − Hc is + or −. (+7 nautical miles)

d. Through this point found in Step **c** draw a line of position at right angles to the bearing Zn.

Labeling of LOP: The line of position (LOP) should be correctly labeled, with the time of sight *above* the line, and the name of the celestial body *beneath* the line.

Step 12: The corrections necessary to reduce Hs to Ho were listed in Part III with explanations.

UNIVERSAL PLOTTING SHEET; LONGITUDE SCALE. For Requirement 2, note that on a Universal Plotting Sheet the parallels of latitude are already drawn, and you must draw the necessary meridians of longitude. Since a universal plotting sheet is really a simple Mercator chart without expanding latitude on one sheet, the longitude scale in the lower right-hand corner shows the length of 1° of longitude. Take dividers, place each point at opposite ends of the appropriate mid latitude line (43° in this Requirement). Mark off degrees of longitude as shown on Plotting Sheet No. 1. Note further that one interval of 10′ of longitude is divided into five parts. Each part is one-fifth of 10′ or 2′. These parts enable you to measure different, smaller values of longitude at the mid latitude, with the aid of the dividers. Construction of this chart is explained in the author's *Coastwise Navigation* and also *Particularized Navigation.*

CHECKING THE TIME AT A KNOWN POSITION. To interrupt the Practice Cruise, if he is interested, before leaving this morning sun sight of the Practice Cruise, the junior navigator may follow here the detailed steps suggested in Part VI, Section W under Checking Time at Known Position. We could assume, for instance, that at the time of his morning sight the navigator was unsure of the accuracy of his watch. Let us assume, for example, that his watch read 10^h56^m00 GMT at time of the sextant sight of the sun, and that the navigator was absolutely sure of his position then, at 42°53′N, 29°40′W. He now wants to go into reverse operation and check the accuracy of the watch. The steps are as follows:

Ho = 40°39.8′ (sextant sight of sun)
With GMT 10ʰ56ᵐ sun's declination is 15°05.2′N
$$\frac{-0.7}{15°04.5'N}$$

Declination (north) is *same* name as latitude. On the Same Name page of *Pub. No. 249* for Latitude 43°, and in the declination column headed by 15° (next below declination of 15°04.5′), the junior navigator will note that the tabulated Hc of 40°29′ is *closest* value to his Ho of 40°39.8′ and also *smaller* than the Ho. He knows that for Declination 15°04.5′ the tabulated Hc will be 4′ larger than 40°29′ ($\frac{4.5}{60} \times d$ value of 42 equals 4′, from Table of Correction to Tabulated Altitude for Minutes of Declination). Hence, for the declination 15°04.5′N the correct tabulated Hc = 40°33′, while the Z = 109° and LHA = 312° as read from *Pub. No. 249* in the row which contains 40°29′. Ho − Hc = 40°40.0′ − 40°33′ = +7′ = 7 nautical miles.

If the navigator plots Zn of 109° through his *known position* 42°53′N, 29°40′W, and then measures 7 nautical miles *away* from the sun, he will see that the assumed position for the *Pub. No. 249* Tables is 43°N and 29°43.7′W. (As one check, the navigator knows that the assumed position would be at 43°N.)

The remainder of the problem is easy:
LHA + long. W. = GHA
LHA of sun = 312°
$$\underline{29\ 43.7'W}$$
GHA of sun = 341°43.7′

Turning next to the *Nautical Almanac* for August 12 the navigator asks "What GMT gives 341°43.7′ for sun's GHA?" At 10ʰ GMT on August 12, sun's GHA = 328°43.7′. What minutes and seconds will give 341°43.7′? There is a difference of 341°43.7′ − 328°43.7′ or 13°00.0′. Turning to the *yellow page* of 52ᵐ in the *Nautical Almanac*, the junior navigator notes that 13° = 52ᵐ.

Hence GMT 10ʰ52ᵐ00ˢ (to the nearest minute) is the correct time to give the Ho of the sun at the assumed known position. The navigator then sees that his watch is fast by 10ʰ56ᵐ00ˢ − 10ʰ52ᵐ or by 4ᵐ. If this position, for instance, had been at a known island, the navigator could have

checked his watch just as Worsley did on Elephant Island when he was about to leave the island with Shackleton on the long boat journey to South Georgia Island.

CORRECTION TABLE FOR MOTION OF VESSEL WITH FAST SPEED. For Requirement (3) note the statement just above the Table of Corr. to *t* on the Form for Reduction of Noon Sight, p. 97. For practical purposes an error of 10′ in the value of the correction will lead to an error of only 40ˢ in the Predicted Watch Time of Noon. If the speed of the vessel is less than 12 knots, it is not necessary to apply a correction to *t* in order to obtain an accurate latitude at meridian passage of the sun. In other words, most yachtsmen of today may ignore the Table of Corr. to *t*. For the S. S. *Ruchbah* a correction is necessary. (See Part IV, where the details of this reduction are carried out, for theory and further explanation of this noon sight; also see filled-in Form, Part X, p. 97.)

NOON CURVE. For Requirement (5) the scale is most important in drawing a noon curve. It can be too small, or it can be too large for an accurate drawing of the noon curve. Examine the scale in Fig. 22, where it is about right.

For Requirement (6) note that an accurate latitude does not depend upon an exact time. The time could be about 3ᵐ in error. The latitude, however, will be accurate; it can be accurate to within 0.2 nautical miles after practice with sextant sights.

For Requirement (7) note that the longitude from the noon curve, 28°11′W, differs by almost 4′ from the longitude as derived in Requirement (9): Longitude 28°07′W from a noon running fix at 1358. This difference is equivalent to about 3 nautical miles at this latitude. The error in determination of longitude from a noon curve is apt to run from about 3 to 10 nautical miles. It might even be as great as 20 nautical miles for a student navigator who has not had much practice in drawing a smooth noon curve through his sextant sights.

RUNNING FIX FROM SUN. For Requirement (9), the LOP from the 1052 sun sight is advanced along the course just as a line of position in piloting is advanced. From the intersection of the 1052 LOP and true course, mark off a DR distance along the course for the time difference (1358-1052) and speed of 22 knots. At the end of this distance draw a second line (broken for identity) parallel to 1052 LOP. Label latter LOP with the two sextant-sight times. Intersection of this advanced LOP with a second LOP at 1358 gives what is termed a "running fix." If the course between the two times is quite steady over the bottom, the running fix can be accurate; of course, with both wind and current often unknown, more uncertainty should be attached to a running fix. However, during the day, with no other celestial body in the sky, running fixes from the sun can be of great value, and should be obtained if nothing better is offered. Sun sights certainly proved of the greatest value to Worsley in the *James Caird* along the Drake Passage, as mentioned in Part III.

Data from *Nautical Almanac* for Moon Sight Reduction

v AND *d* CORRECTIONS FOR MOON. For Requirement (11), in reducing a sextant sight of the moon, there is an additional correction (*v* correction) which must be added to the GHA of the moon. The hourly rate of increase of the GHA of the moon is not as regular as it is for the sun and for the first point of Aries. Hence this additional correction is necessary. Note the *v* value given for the moon on the daily page of the *Nautical Almanac*. Proceed as you would for *d* corrections for declinations of the sun. Find the *v* value on the appropriate *yellow page* of the *Nautical Almanac* (according to the minutes of the GMT of the sight). Read the *v* correction, just as a *d* correction is read, and always *add* this value to the GHA of the moon. For the *d* correction the sign of the correction is always determined by *inspection* of the declination values for the hour preceding and the hour following the time of observation. The *v* correction sign for the moon is always +.

ALTITUDE CORRECTION FOR MOON. For reducing the Hs of the moon to the Ho, the special Altitude Correction Tables for the Moon at the *end* of the *Nautical Almanac* must be used, mainly because of the large parallax-correction which is important for the moon only, due to its close proximity to the earth as compared with the other bodies used in celestial navigation.

The directions near the end cover page of the *Nautical Almanac* for the moon and headed "Moon Correction Table" (p. xxxiv) can be followed by the student navigator. H. P. (Horizontal Parallax) is given on the daily page of the *Nautical Almanac* in the moon column with H. P. at the top of the page. It is needed on the cover pages for the moon so that the navigator will know in which row to read the added, second correction.

There are separate second corrections; one for a *lower limb*, and the other for an *upper limb* sextant sight, as explained in the directions on this cover page.

VALUE OF MOON SIGHT. It will be noted that for the moon there are more corrections to be applied, and secondly, some of these corrections are larger than they are for the sun. As a result a mistake in arithmetic is more serious in use of the moon than in the use of the sun, stars, or planets. This is probably the reason why many navigators have not enjoyed taking moon sights. Sometimes a feeling of dislike for the moon arose. On the other hand, the value of the moon should be appreciated by every yachtsman and navigator, mainly because of the familiar explanation "It is there!" It is a second celestial body, which can provide a second LOP to go with a sun LOP during any day when the moon is above the horizon. In other words, on many days it can often provide a good fix. Statistically, from the ocean surface the moon and sun offer a better chance for sextant sights than the stars and planets. The former can often be seen through half-covered or thinly covered, cloudy skies. Often, when the stars and planets are invisible at twilight because of clouds, the moon will be the *one* celestial body available to the navigator. Mixter, in his *Primer of Navigation* even claimed that it is sometimes possible to take sights of the full moon at night, when the horizon is illumined by the moon itself. The author has not been too successful with similar sights of the moon, but again it should be emphasized that even an inaccurate sextant sight is better than no sight at all.

UPPER LIMB SIGHT OF MOON. Note, finally, in the moon sight, that the *upper* limb had to be used. Because of the moon's phases, the whole half of the moon can be seen by the navigator only at full moon. Most of the time some of this half of the moon visible to the navigator is dark because it is not receiving light from the sun. The navigator needs to use the illumined limb of the moon, and this may be the upper limb or the lower limb, according to the moon's declination and its general position as related to the sun and to the navigator's horizon. In the *Ruchbah* cruise, for instance, the *upper* limb would be used by the navigator. The lower limb could not be seen clearly.

Data from *Nautical Almanac* for Star Sight Reductions

STAR FIX. For Requirement (14), when possible, three star sights will give an excellent fix. They are possible, as for every celestial body, only when the celestial body and the horizon can be seen simultaneously. The stars can be seen only after sunset and before sunrise. If it is too dark, however, the horizon cannot be seen.

TIME FOR SEXTANT SIGHTS OF STARS. The only times for simultaneous views of stars and horizon are during the moments of twilight, about half an hour after sunset or half an hour before sunrise. These appropriate twilight moments (LMT) are indicated in the *Nautical Almanac* on each daily page under the title of "Twilight, Civil." By definition, civil twilight is the time when the sun is 96° from the zenith. and at this moment the horizon is still clearly visible while it is dark enough to see the brightest stars. At nautical twilight time, also given in the *Nautical Almanac,* it is too dark to observe the horizon. As a result, usually the navigator has *only* about a half-hour of twilight in which to observe three stars and possibly a planet. A bright planet is easier to see than a fainter star, but Vol. 1 of *Pub. No. 249* is designed for reduction of stars only. Hence, for a quick reduction the navigator is apt to choose three stars whenever possible, and use a planet when clouds interfere with some of the star sights. Often, too, the bright planet Venus can be seen long before the stars are visible after sunset. A valuable improvement to a sextant would be a night-vision image intensification telescope, so that the horizon could be seen throughout the night. (See Seidelmann and Feldman in References.)

PRECOMPUTATION OF STAR SIGHTS. One clear fact remains; the yachtsman or navigator does not have time to be too leisurely about his twilight sights of stars. He will save time if during the afternoon he precomputes the Hc of each star which he might use at civil twilight. Seven possible stars are given in *Pub. No. 249* for any particular time or place. Vol. 1 of *Pub. No. 249* is entered with the appropriate assumed latitude and also with the LHA ♈ . The star-sight procedure for the navigator follows:

1. Find, in *Nautical Almanac,* the time of civil twilight. The time given there is LMT. The navigator can obtain the corresponding GMT by applying his DR longitude expressed in time (as explained in Part VI).

2. Find, for this GMT, the corresponding GHA ♈ by turning to the daily page of the *Nautical Almanac* with the hours of GMT. Turn to the *yellow pages* for the minutes and seconds of GMT to find the values of GHA to add to the first value of GHA for hours. The resulting sum equals the GHA ♈ for civil twilight.

3. Find the corresponding LHA ♈ by applying the DR longitude, but with the minutes changed to give a *whole* number of degrees for LHA ♈ .

4. On appropriate page of *Pub. No. 249*, Vol. 1, for an assumed latitude as close to the DR latitude as possible, but equal to a *whole* number of degrees of latitude, find LHA ♈ equal to the LHA of **Step 3**.

5. For LHA ♈ in **Step 4** note the seven stars listed in same row as most useful at this time, and write down the Hc and the Zn for *at least* three of these possible stars which you might choose. Choice will be according to personal preference.

"ASTERISK" STARS IN *PUB NO. 249*, VOL. 1. The stars which give the best intersecting azimuths are marked with asterisks and will be chosen by some navigators. These three will give the best fix as far as the intersecting lines of positions from the three stars are concerned, and they will also minimize possible refraction due to a non-standard atmosphere. The author prefers the brightest stars, especially in partly cloudy skies, and is apt to choose the three brightest stars which also give reasonable intersections. If the azimuth lines of two stars should be parallel, for instance, only one of these could be used. Again, if the angle of intersection of the azimuth lines is quite small, the two lines of positions from these two stars would be as bad as two parallel lines. In other words, the angles between the azimuths must be considered.

Zn IN *PUB. NO. 249,* VOL. 1. It should be emphasized that the azimuths or Zn values given in Vol. 1 of *Pub. No. 249* are the *true azimuths to be plotted.* At the appropriate time of twilight the navigator can face the direction of the azimuth of one of the stars to be used. (The magnetic compass can be used with any necessary correction for variation and deviation.)

IDENTIFICATION OF STARS BY *PUB. NO. 249,* VOL. 1. If the navigator has set on his sextant the Hc given in *Pub. No. 249* for that approximate time, he has only to look at the horizon through his sextant horizon glass and he will see in the field both the horizon and the star. (This is the only star identification used by many navigators.) The navigator will then turn the micrometer drum until the star is tangent to the horizon as he *rocks* his sextant. At this moment the navigator will call "mark," either to an assistant or to himself. He will note the corresponding time of the sextant sight, counting seconds until he can read the time on his watch, if alone, and finally he will read the Hs, on the sextant arc, for this star. With practice the navigator can take three or even four sextant sights of each of the three stars. He will plot the three Hs's for each star, to make sure that they lie on a straight line, so that any one of the three is accurate and may be reduced. In this way the navigator practices constant vigilance. However, if the weather is cloudy, with danger of losing a star in the clouds, the navigator may decide to take single sextant sights on each star. If the weather permits he can still repeat the round of star sights, even twice. At least he is sure of obtaining one fix from the first three star sights.

STAR FIX OBTAINED IN 15 MINUTES. The sextant sights may be reduced and the fix obtained in only about 15 minutes (after practice) if *Pub. No. 249* is used. The details may be followed on the Star Fix (*Pub. No. 249*) Form in Part X, p. 100.

STAR FIX FROM 4^m-INTERVALS. After following these details it is easy to understand how the reduction time can be shortened further by taking three star sights exactly four minutes apart. If this exact timing can be achieved with the sextant sights of stars, only one complete reduction is necessary—the one for the middle star sight, as seen on the Star Fix (*Pub. No. 249*) Form No. 7 for 4^m Time Intervals, p. 101. Because of the 4^m interval, the LHA ♈ for the preceding star will be 1° less, and the LHA ♈ for the following star 1° more in LHA ♈ than the computed LHA ♈ for the middle sight. (It should be remembered that 1° = 4^m.) These corresponding values can be entered on the Form as shown. They can then be used to enter *Pub. No. 249,* Vol. 1 with the appropriate latitude, stars, and LHA ♈. The corresponding Hc and azimuth for each star can be entered on the Star Fix Form, so that the intercepts may be found. In Part X, for comparison, a reduction for 4^m intervals of the same three stars is presented on p. 101.

The plotting of the three lines of position follow as for a sun reduction. There is just one further correction to consider, due to the fact that the declination values for the three stars were incorporated and considered by the *Nautical Almanac* Office in deducing the Hc and azimuth for each star. In his use of *Pub. No. 249,* Vol. 1 one might think that the navigator could ignore the declinations of the stars, but they cannot be ignored altogether.

PRECESSION AND NUTATION CORRECTION. During one year the declination of a star changes only slightly (by tenths of a minute of arc), but after a few years the cumulative change in declination for a star cannot possibly be ignored by the navigator. *Pub. No. 249* handles the situation by supplying a correction table called "Precession and Nutation Correction." This Table is entered with the appropriate year, the appropriate latitude and with the LHA ♈ at the time of the observation.

APPLYING CORRECTION TO FIX. After a fix is obtained, it can be moved the number of nautical miles indicated in the table, and in the direction indicated. 1'100°, for instance, tells the navigator to move his fix one nautical mile in the true direction of 100°. Every few years a new edition of *Pub. No. 249,* Vol. 1 is printed for a certain year or epoch. *The Extract given in Part X is for the year or epoch 1980.* Corrections for 1981 could almost be ignored, since they are equal to 1′ at most. For 1982 the corrections run from 0 to 3′ depending upon latitude and LHA ♈, and in 1984 they will sometimes amount to 4′. Before 1985 a new edition of *Pub. No. 249,* Vol. 1 will be available. The precession and nutation problem does not affect use of *Pub. No. 249,* Vols. 2 and 3.

ADVANCING LOP FOR STARS. Depending upon accuracy of sextant sight, speed of the vessel, and interval of time between sights, a line of position from a star at an earlier time should be advanced just as a line of position in piloting is advanced. It should be advanced along the true course of the vessel to go with the line of position at a later time as described above in the discussion of Requirement (9). Star sights are usually close in time and may not require advancing of an LOP except for high speeds. *Retiring* is advancing in opposite direction.

SIGHT ON STAR ABEAM TO COURSE. It should be noted that if a celestial body is *abeam* to the vessel, its line of position will be parallel to the true course of the vessel. It will never need to be advanced to go with a later sight as long as the course is not changed. This is an advantage to be remembered when possible stars are chosen for sextant sights.

As a final supplementary remark on star sights, constant vigilance is observed more faithfully if the navigator takes some time to learn the identity of the brightest navigation stars. For greatest safety he should not depend upon any outside aid for identification. If, however, a bright star is in a favorable position at any time or place, with a chance of its disappearance into the clouds shortly, it can be sighted first and identified later if necessary. No time should be lost.

STAR CHARTS. A *Star Finder* (formerly H. O. 2102-D, now 1H-6606-129-6526) has reassured some student navigators about star identity, but daily use of *Pub. No. 249,* Vol. 1 on clear nights can be very effective along with study of the "Star Charts" of the *Nautical Almanac,* pp. 266, 267. The circular polar charts should be held out with the celestial pole (center) of the chart at the altitude of the DR latitude approximately. The chart can be oriented by the azimuth of one of the brightest stars given in *Pub. No. 249,* Vol. 1 for some particular time. If, for instance, Arcturus was shown to have an azimuth of 100°, the Star

Chart can be revolved until Arcturus is seen to be in that direction, 10° to the south of east on the horizon. True north, south, and west can then be noted and the navigator should be able to inspect any constellation or group of stars (indicated on the chart by capital letters). Looking from the chart to the celestial sphere, he can find the corresponding stars in the sky. He should remember that in a navigator's lifetime the stars will keep the same relation to each other, and distances between them will be maintained. He can note the declination figures given on the charts for a scale or distance of 20°, and with practice the navigator can travel from a known star or constellation, such as URSA MAJOR, to another star nearby. It is best to travel a short distance each time, stepping in imagination from one constellation to the next. The celestial sphere has been projected onto a plane with some distortion if one travels too far (more than 50° for instance). The circular Star Charts extend only to a declination of 10°, and "Equatorial Stars" are pictured in rectangles at the bottom of the same pages of the *Nautical Almanac* to take care of constellations and stars along the celestial equator. The SHA (Sidereal Hour Angle) figures shown may be an aid in traveling from a circular chart to the equatorial chart. A star near SHA 120° on the circular chart will also be near SHA 120° on the equatorial chart. Note that SHA increases toward the west, just as LHA and GHA do. With *Pub. No. 249,* Vol. 1, many navigators no longer need to use SHA, which is shown on Form 1 for stars not given in *Pub. No. 249,* Vol. 1.

REDUCTION OF PLANET SIGHT. For Requirement (22) the reduction of a planet sight is similar to the reduction of a sun sight. There are two additional corrections involved. One is a *v* value and a *v* Corr^n. similar to these *v* values for the moon. Again this correction is necessary because the planets have slight irregularities in GHA motion. In the case of Venus this *v* value sometimes has a minus sign preceding it, indicating subtraction. The *v* value is found at the bottom of the page, beneath the GHA column on the daily page of the *Nautical Almanac,* and the corresponding *v* Corr^n. is found on the *yellow page* which has the minutes of the GMT sight at the top. The other additional correction for the planet Mars, and also for the planet Venus, is shown on the cover page of the *Nautical Alma-*

nac in the middle of the page, under the heading "Stars and Planets." This correction is applied to the Hs of the planet. Mars and Venus are the planets closest to the earth; hence corrections for parallax and phase (similar to phase of the moon) provide greater accuracy.

IDENTITY OF PLANETS. In general, "Planet Notes," p. 8 of the *Nautical Almanac,* describe the positions of the planets for each year. Positions change from year to year since the planets have different periods of revolution around the sun. Because of their nearness as compared with stars, the four planets used in celestial navigation, Venus, Mars, Jupiter, and Saturn, appear as disks through binoculars. Each whole disk of a planet is not affected by variation in refraction as much as the apparent ray of light from a distant star. Refraction of a stellar ray varies because of changing density in the earth's atmosphere, while many rays from a disk usually are not displaced in the same direction at the same time. Hence, the planets do not twinkle as much as the stars. This fact often enables a navigator to distinguish between a star and a planet, especially when the two are close together in direction, when the need to distinguish is greater than it normally is. In addition, times of meridian passage (LMT) are given for the planets, just as they are for the sun, on the daily pages of the *Nautical Almanac.* (See Part VI.) The LMT of meridian passage enables the navigator to identify the planets if he also remembers that the approximate rate of apparent motion on the celestial sphere is always 15° per hour. Whenever the need arises, he can roughly estimate the distance in degrees of each planet from the celestial meridian, upper branch. Finally, it is important to know that the planets, as well as the moon, lie approximately along the ecliptic, almost through Regulus and Zubenelgenubi, as shown on the Equatorial Star Chart in the *Nautical Almanac.* Some navigators might prefer to use Spica rather than Zubenelgenubi as a guide star for the ecliptic. Spica is a brighter star with a shorter name.

PART IX

Use of the True Azimuths of the Sun (From *Pub. No. 260* or *Pub. No. 249*) as a Check on the True Headings of a Vessel

The yachtsman with a new or uncompensated compass, or the experienced navigator in an emergency may not know how to compensate the magnetic compass. They may not be sure that the magnetic compass has been compensated; but if they are wise, they will want to check the deviation (error) card for the compass, or at least check the headings one by one as they cruise. They do not have to worry about any errors in magnetic headings if they wait for fair weather and use the sun and a book of tables (either *Pub. No. 260* or *Pub. No. 249*).

Pub. No. 260, once known as "Red Azimuth Tables," gives the azimuth of the sun for every ten minutes during the day.

X
Use of *Pub. No. 260* for True Azimuths of Sun

Suppose, for example, the yachtsman reads from *Pub. No. 260* that the azimuth of the sun is 130° at a certain time in the morning of a given day. He can then set the vanes of his pelorus so that the sun has a true bearing (or azimuth) of 130°. All other directions on the pelorus are now true directions and the true heading of the vessel can be checked. This is the most accurate method.

USE OF PROTRACTOR AS A PELORUS. For very approximate accuracy, if the yachtsman lacks a pelorus, in an emergency he may use a simple plastic protractor and his own hand as a substitute for the vanes of his pelorus, as shown in Fig. 36. The center of the protractor, which is functioning as a pelorus, should lie along the keel of the vessel preferably. It might also lie on a seam between the wooden planks which run parallel to the keel or on some similar line running parallel to the keel. The yachtsman can hold his hand out as flat as possible, to simulate the vanes of a pelorus. He can place his hand over the center of the protractor and turn the protractor under his hand until he has lined up 130° on the protractor, his hand, and the sun as seen over the hand, all in the same direction of 130°. (To avoid any injury to the eye it is best to "peek" at the sun over the hand.) All other directions on the pelorus, or protractor, are then true directions, measured from the north (0°) of the protractor, which is now the direction of the true geographical North Pole on the earth. If the keel lies in the direction of 170° and toward the bow as indicated by the protractor, the true heading is 170°, and this true heading of 170° is the correct true course to be plotted on a chart. The error in any true direction determined in this simple way may be about 5° (or possibly a few degrees more) depending on the plumpness of the hand!

SIZE OF PROTRACTOR. One protractor of about 4 inches in diameter should always be carried in the yachtsman's pocket for an emergency, while another larger one, of *about 7 inches,* should be used for greater accuracy.

Any change in course should be checked for true direction in similar fashion, since the magnetic compass may have different errors (or deviations) on different headings.

68

VARIATION. As the yachtsman probably knows, besides a possible deviation, a second correction must be considered: the angular difference between two directions, true north and magnetic north, is called Variation, which de-steer 170° + 15° or 185° by a compensated magnetic compass (with no deviation). In another position on the earth, if the variation equaled 15°E, he would steer 170° − 15°, or 155° by the magnetic compass. The use of a pelorus, real or

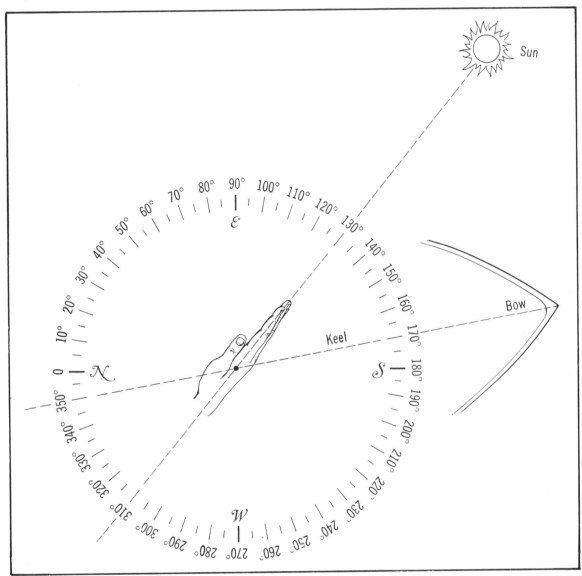

Fig. 36. Protractor and mariner's hand in simulation of a pelorus—for an emergency.

pends upon the geographical position of the vessel. The Variation can be found on all ocean and coastal charts: Variations are also shown on pilot charts of the oceans, published monthly. A navigator always carries a pilot chart for the appropriate ocean and month. (DMA PILOT 168107 for the North Atlantic), the scene of the Practice Cruise of Part VII.

EXAMPLE IN VARIATION. If, for example, the variation equals 15°W and the yachtsman wanted to steer a true course of 170°, he would

simulated by "hand," would check the heading as explained above, or would indicate the heading, or even discover it in case the yachtsman were lost. In other words, use of *Pub. No. 260* and of true azimuths of the sun takes care of both variation and deviation at the same time, and checks the true heading which the yachtsman or navigator may wish to plot on his chart. If he prefers, he may plot magnetic courses, but he must remember that the azimuths in *Pub. No. 260* or *Pub. No. 249* are true directions, from true north.

TRUE BEARING OR AZIMUTH.

LATITUDE 41°

DECLINATION—SAME NAME AS—LATITUDE.

Dec.	12°	13°	14°	15°	16°	17°	18°	19°	20°	21°	22°	23°	Dec.

Apparent Time A. M.

Month	12°	13°	14°	15°	16°	17°	18°	19°	20°	21°	22°	23°
April / May / June	22	25	28	1	5	8	12	16	21	26	1	10
August / July	22	19	16	12	9	5	2	28	24	19	12	3
October / November / December	25	28	31	3	6	10	14	17	22	27	3	11
February / January	18	15	12	9	5	2	29	25	21	16	10	2

Apparent Time P. M.

A.M. h.m.	12°	13°	14°	15°	16°	17°	18°	19°	20°	21°	22°	23°	P.M. h.m.
IV 40											60 36	59 54	20
50									63 40	62 57	62 14	61 31	10
V 00						67 29	66 45	66 01	65 17	64 34	63 50	63 06	VII 00
10				70 35	69 50	69 06	68 22	67 38	66 53	66 09	65 24	64 40	50
20	74 26	73 41	72 57	72 12	71 28	70 43	69 58	69 13	68 28	67 43	66 58	66 12	40
30	76 04	75 19	74 34	73 49	73 03	72 18	71 32	70 47	70 01	69 16	68 30	67 44	30
40	77 41	76 55	76 10	75 24	74 39	73 53	73 07	72 21	71 35	70 48	70 01	69 15	20
50	79 17	78 31	77 45	76 59	76 13	75 27	74 40	73 54	73 07	72 19	71 32	70 45	10
VI 00	80 53	80 07	79 21	78 34	77 47	77 00	76 13	75 26	74 38	73 51	73 03	72 14	VI 00
10	82 29	81 42	80 55	80 08	79 21	78 34	77 46	76 58	76 10	75 21	74 32	73 43	50
20	84 05	83 18	82 31	81 43	80 55	80 07	79 19	78 30	77 41	76 52	76 02	75 12	40
30	85 41	84 54	84 06	83 18	82 28	81 40	80 51	80 02	79 12	78 22	77 32	76 41	30
40	87 18	86 30	85 41	84 53	84 04	83 14	82 24	81 34	80 42	79 53	79 01	78 10	20
50	88 55	88 06	87 17	86 28	85 39	84 48	83 58	83 07	82 16	81 23	80 31	79 39	10
VII 00	90 33	89 44	88 55	88 05	87 14	86 24	85 32	84 40	83 48	82 55	82 02	81 08	V 00
10	92 13	91 23	90 33	89 42	88 51	87 59	87 07	86 15	85 22	84 28	83 33	82 39	50
20	93 54	93 03	92 12	91 21	90 29	89 37	88 44	87 51	86 56	86 01	85 05	84 10	40
30	95 36	94 45	93 54	93 02	92 09	91 16	90 22	89 27	88 32	87 36	86 40	85 43	30
40	97 21	96 29	95 37	94 44	93 50	92 56	92 01	91 06	90 10	89 13	88 15	87 17	20
50	99 08	98 15	97 22	96 29	95 34	94 39	93 43	92 47	91 49	90 51	89 52	88 52	10
VIII 00	100 57	100 04	99 10	98 16	97 21	96 25	95 28	94 30	93 31	92 32	91 32	90 30	IV 00
10	102 50	101 56	101 02	100 06	99 10	98 13	97 15	96 16	95 16	94 15	93 14	92 11	50
20	104 46	103 51	102 56	102 00	101 03	100 05	99 05	98 05	97 04	96 02	94 59	93 54	40
30	106 46	105 51	104 54	103 57	102 59	102 00	101 00	99 58	98 56	97 52	96 47	95 41	30
40	108 50	107 54	106 57	105 59	105 00	104 00	102 58	101 52	100 52	99 47	98 40	97 32	20
50	110 59	110 03	109 05	108 06	107 07	106 05	105 02	103 58	102 53	101 46	100 37	99 28	10
IX 00	113 14	112 17	111 18	110 19	109 17	108 15	107 11	106 06	104 59	103 50	102 40	101 29	III 00
10	115 34	114 37	113 38	112 37	111 35	110 32	109 27	108 20	107 12	106 02	104 49	103 36	50
20	118 02	117 03	116 04	115 03	114 00	112 55	111 49	110 41	109 31	108 19	107 05	105 49	40
30	120 36	119 38	118 38	117 36	116 33	115 27	114 20	113 11	111 59	110 46	109 30	108 12	30
40	123 19	122 20	121 20	120 18	119 14	118 08	117 00	115 49	114 37	113 22	112 04	110 44	20
50	126 11	125 12	124 12	123 10	122 05	120 59	119 50	118 38	117 24	116 07	114 49	113 26	10
X 00	129 12	128 14	127 14	126 12	125 08	124 01	122 51	121 39	120 25	119 07	117 46	116 22	II 00
10	132 23	131 27	130 27	129 26	128 22	127 15	126 06	124 54	123 38	122 20	120 58	119 32	50
20	135 46	134 51	133 53	132 52	131 49	130 44	129 35	128 23	127 07	125 48	124 26	122 59	40
30	139 21	138 27	137 31	136 33	135 31	134 27	133 19	132 08	130 54	129 35	128 13	126 45	30
40	143 08	142 17	141 24	140 28	139 28	138 26	137 21	136 12	135 00	133 42	132 20	130 53	20
50	147 08	146 20	145 30	144 37	143 42	142 43	141 41	140 35	139 25	138 10	136 51	135 26	10
XI 00	151 19	150 37	149 51	149 03	148 11	147 17	146 20	145 18	144 13	143 03	141 47	140 27	I 00
10	155 44	155 06	154 26	153 43	152 57	152 09	151 18	150 23	149 23	148 20	147 11	145 57	50
20	160 20	159 48	159 14	158 38	158 00	157 19	156 35	155 47	154 57	154 03	153 02	151 56	40
30	165 05	164 40	164 14	163 46	163 16	162 43	162 09	161 31	160 51	160 06	159 18	158 25	30
40	169 59	169 42	169 24	169 04	168 44	168 21	167 57	167 31	167 03	166 32	165 58	165 20	20
XI 50	174 58	174 49	174 40	174 30	174 20	174 08	173 56	173 42	173 28	173 12	172 54	172 35	XII 10

	12°	13°	14°	15°	16°	17°	18°	19°	20°	21°	22°	23°	
Sun rises	5 17	5 14	5 10	5 06	5 02	4 58	4 54	4 50	4 46	4 42	4 38	4 33	Sun rises
Sun sets	6 43	6 46	6 50	6 54	6 58	7 02	7 06	7 10	7 14	7 18	7 22	7 27	Sun sets
Azimuth	74 01	72 40	71 18	69 57	68 35	67 12	65 50	64 27	63 03	61 39	60 14	58 49	Azimuth

In **North latitude**, when the body is rising or East of the meridian, the tabulated azimuths are reckoned from North to East; and when the body is setting or West of the meridian, the tabulated azimuths are reckoned from North to West.

In **South latitude**, when the body is rising or East of the meridian, the tabulated azimuths are reckoned from South to East; and when the body is setting or West of the meridian, the tabulated azimuths are reckoned from South to West.

READING *PUB. NO. 260*. The reading of azimuths of the sun from *Pub. No. 260* must now be investigated. The accuracy necessary in reading the azimuth of the sun depends somewhat on the steering accuracy. If a boat can be steered only with a possible error of 5°, the yachtsman does not have to worry too much about the kind of time he uses in entering *Pub. No. 260*. The time used in *Pub. No. 260* is local apparent time (LAT). If the steering error is 5° or more, the yachtsman may use local mean time (LMT) as if it were local apparent time. To obtain the local mean time for entering *Pub. No. 260,* it is easiest if the yachtsman carries GMT on his watch, as suggested in Section W. He obtains the corresponding local mean time by subtracting or by adding the DR longitude. For example, if GMT = 15^h00^m, April 26, at DR Lat. 41°10'N, Long. 68°40'W, the yachtsman applies his DR longitude (expressed in time) to the GMT. He subtracts the longitude, since the latter is west. The computation is as follows:

$$GMT = 15^h00^m00^s \text{ April 26}$$

Long. W = $\underline{\quad 4 \; 34 \; 40 \quad}$ Arc to Time Table, first yellow page of *Nautical Almanac* (see p. 79 of this book)

$$LMT = 10 \; 25 \; 20 \text{ April 26}$$

ENTERING *PUB. NO. 260*. The yachtsman next turns to the latitude page in *Pub. No. 260* which is closest to his latitude, and where, also, Declination Same Name as Latitude is printed at the top of the page, in Part II. Part III takes care of Declinaton Contrary Name to Latitude, while on the one page of Part I the declination may be north or south. For the problem above, the yachtsman would turn to the page of Part II with 41° at the top. (See p. 70.) He reads the azimuth value in the column headed April 25, the date closest to his given date, April 26. In the left (A.M.) column, for 10^h20^m (X^h20^m) he reads 135°; for 10^h30^m (X^h30^m) he reads 138° (to nearest degree). For LMT 10^h25^m the azimuth is 137°. If, however, at the same DR and on the same day, April 26, the GMT = 20^h00^m, five hours later, in the afternoon, the figures for LMT would run as follows:

$$GMT = 20^h00^m00^s \text{ April 26}$$

Long. W = $\underline{\quad 4 \; 34 \; 40 \quad}$

$$LMT = 15 \; 25 \; 20 \text{ April 26}$$

Since the local time is in the afternoon, the figures on the *right* side of the page should be used, as indicated.

RULES FOR P.M. AZIMUTHS. In the April 25 azimuth column, for 03^h20^m (III^h20^m) P.M., the yachtsman reads 108°; for 03^h30^m (III^h30^m) he reads 106°, to the nearest degree. For $15^h25^m20^s$, halfway between in time, he finds 107° to be the nearest table value. Next, the fine print at the *bottom* of each page of *Pub. No. 260* is most important to remember. In other words, in the afternoon the figure in the azimuth column does not give the direct azimuth, or the angle plotted from the north clockwise. In this example, according to the fine print, 107° must be subtracted from 360° since the tabulated azimuth is reckoned from north to west when the sun is west of the meridian, in the afternoon. The azimuth to be plotted on the chart is 360° − 107° = 253° (from true north).

RULES FOR SOUTH LATITUDE. Whenever the latitude is south, as stated in the fine print at the bottom of the page of *Pub. No. 260,* in case the sun is east of the meridian, in the morning, the tabulated azimuth in the table must be subtracted from 180°. The result is the true azimuth which gives the true direction from north. In the afternoon, on the other hand, when the sun is west of the meridian, 180° is added to the tabulated azimuth in the table in order to obtain the true azimuth from north. With these rules the tables in *Pub. No. 260* can be more condensed than they would be if only true azimuths from north were given in all cases. They are never out of date, and that explains why H. O. can be retained in the title.

USING *PUB. NO. 260* **WITH GREATEST ACCURACY.** For the navigator who wishes it, more accurate interpolation may be carried out in the use of *Pub. No. 260*. Also, local apparent time may be used. LAT may be obtained easily from LMT by applying the equation of time to the LMT. The equation of time is defined in Part VI. (see p. 49) It is found on the lower right-hand page of the daily page of the *Nautical Almanac,* for 00^h GMT and also for 12^h GMT on each day of the year. Interpolation may be applied to obtain the equation of time to the nearest second, but only the nearest minute is necessary for use in *Pub. No. 260*. One problem always arises at this point: "Is the equation of time added or subtracted from the LMT to obtain the LAT?" Inspection, on the same daily page, of the "Mer. Pass." column gives the LMT when the sun crosses the upper branch of the celestial meridian. Fortunately the LAT of meridian passage is *always* 12^{h}00^m, as was explained more fully in Section T.

FINDING LAT FOR USE IN *PUB. NO. 260*. If the LMT of meridian passage is less than 12^{h}00^m, the equation of time must be *added* to any LMT of that day to obtain the corresponding LAT. If the LMT of meridian passage is greater than 12^{h}00^m, the equation of time must similarly be *subtracted* from the LMT to obtain the LAT. For the example above, on April 26 the equation of time is 2^m, to the nearest minute, and meridian passage is at 1158 LMT. 11^{h}58^m is less than 12^{h}00^m, and hence the equation of time, 2^m in this example, must be added to the LMT. In the above example, at Lat. 41°10′N, Long. 68°40′W,

$$\text{GMT} = 15^h00^m00^s \text{ April 26}$$
$$\text{Long. W} = \underline{4\ 34\ 40}\ \text{Arc to Time table}$$
$$\text{LMT} = 10\ 25\ 20\ \text{April 26}$$

$$\text{LMT} = 10\ 25\ 20\ \text{April 26}$$
$$\text{Eqn. of Time} = \underline{\quad 2 \quad}$$
$$\text{LAT} = 10\ 27\ 20 = 10^h27^m\ \text{to nearest}$$
minute April 26.

In the afternoon, when
$$\text{GMT} = 20^h00^m00^s \text{ April 26}$$
$$\text{Long. W} = \underline{4\ 34\ 40}$$
$$\text{LMT} = 15\ 25\ 20 = 15^h25^m\ \text{April 26}$$
$$\text{Eqn. of time} = \underline{\quad 2 \quad}$$
$$\text{LAT} = 15^h27^m \qquad \text{to nearest minute}$$
April 26

2^m is practically negligible in finding a correct azimuth in this case, but the equation of time may be as large as 16^m. In the latter case, for high latitudes, a 3° error could be obtained by use of LMT instead of LAT for entering the tables of *Pub. No. 260*. The error associated with using LMT instead of LAT is usually negligible for small boats, but it is worth consideration for larger vessels, or if the compass is being "swung" in smooth water. (See author's *Coastwise Navigation*, or Bowditch.)

Y
Use of *Pub. No. 249* for True Azimuths

To check true azimuths and directions, many navigators and yachtsmen prefer the other table, *Pub. No. 249*. The Zn (azimuth to be plotted) of the sun (obtained in the reduction of a sextant sight of the sun at any time during the day, as explained in Parts V, VII, VIII, and IX) could replace the use of *Pub. No. 260*. *Pub. No. 249* gives the altitude and azimuth of the sun for different values of LHA of sun at different latitudes and for different declinations of the sun. For any instant of GMT, which a wise navigator will carry on at least one timepiece, the *Nautical Almanac* gives the declination and GHA of the sun. Adding or subtracting the DR longitude to or from the GHA of sun at the time provides the LHA of sun.

FINDING Zn FROM Z. With declination and LHA of sun on the page closest to the DR latitude, the navigator can read from *Pub. No. 249* the Z (azimuth angle in the navigation triangle, Section Q). At north latitudes Z either equals the Zn (true azimuth to be plotted) or can be subtracted from 360° to find the Zn to be plotted, as stated in Section Q. Directions for going from Z to

Zn are stated on the *Pub. No. 249* Form and at the top and bottom of pages in *Pub. No. 249* Tables.

In summary, *Pub. No. 249* may be used to find the true azimuths of the sun just as conveniently as *Pub. No. 260*. One of these tables may be preferred by one navigator, and the other by another navigator.

It should be emphasized, however, that if a sextant sight is taken and reduced, the true azimuth Zn *has* to be obtained. This Zn can also be used as check on the true heading of the vessel at the moment. *Pub. No. 249* really performs two services for the yachtsman, giving a line of position when a sextant sight has been taken, and also checking the heading and any other true direction. On the other hand, *Pub. No. 260* is concerned with azimuths only, and offers no information on altitudes. Its arrangement, however, might be more pleasing to some yachtsmen. If space is at a premium, *Pub. No. 249* will suffice for two purposes, and *Pub. No. 260* may be left at home. Near shore, for a yachtsman who will never require the altitudes of celestial naviga-

Photograph from Dr. Norris D. Hoyt: checking the magnetic compass for error by observing the azimuth of the sun at an exact moment. This compass has a shadow pin. Ridge White is at the helm.

tion, *Pub. No. 260* can be most useful as a check on the headings. (It is published by DMAHC and may be obtained from the DMA Office of Distribution.)

Z
Importance of Magnetic Compass

In conclusion of this part, the importance of a magnetic compass should be emphasized. It may well be the *only* compass on board many smaller vessels. In this case its deviations on every heading should be known, and can be found either by use of *Pub. No. 260* or by *Pub. No. 249* and the *Nautical Almanac.* On ships with a gyrocompass, in case the latter should fail (due to power failure, for example), the magnetic compass would then become the *only* compass and its deviations should be found or checked as above.

DIRECTIONAL GYRO. Finally, there have been many occasions when the magnetic compass was the best instrument to trust for directions. In the air, for instance, to quote from *H. O. Pub. No. 216:* "a Directional Gyro is an instrument which utilizes the inertia of a gyroscope to provide a reasonably steady reference direction, thus enabling a pilot to maintain a constant heading and to make precise turns. The reading of a directional gyro must be set by reference to some other known reference, as a magnetic compass or the azimuth of a celestial body, but once set, it is independent of all magnetic fields and is not affected by normal amounts of roll, pitch, or yaw of the aircraft. The gyroscope is subject to *wander* and *precession,* and consequently drifts slowly away from the original setting. During the brief interval needed for a turn, the drift is too small to be significant, but if the directional gyro is used to maintain heading over an extended period, it requires resetting rom time to time." In brief, after some interval of time in the air the magnetic compass is apt to give a more reliable direction.

ORION'S BELT (RISING) SHOWS TRUE EAST DIRECTION. Between Aden and Masira, as recounted in "Around the World" (*Flying,* November, 1967), Larry Rockefeller noted a difference of 30° between his magnetic compass and his two directional gyros. He decided to believe his compass, and so escaped contact with the desert or the Indian Ocean! He also knew his stars well enough to remember that Orion's Belt (especially Alnilam and δ Orionis) lies on the celestial equator. Hence, when Orion's Belt is rising (that is, when it is on both the celestial equator and the horizon) it must be at the intersection of these two great circles. It must then be *due east* (Fig. 1). On this last leg of his flight, between Hawaii and San Francisco, Larry Rockefeller realized that he could check his NE true course by noting that he was flying approximately between Polaris (or true north) and Orion's Belt. In other words, knowledge of many relations between stars and constellations can sometimes be a substitute for the use of *Pub. No. 260* and *Pub. No. 249,* especially in regard to true directions or azimuths.

NEED FOR CONSTANT VIGILANCE. Constant vigilance should always be practiced, as it was by Chichester, Rockefeller, Tabarly, Worsley, and all other successful navigators. Spare equipment should always be taken whenever possible, and concentrated attention constantly given to avoidance of blunders and errors. With this concern for *constant vigilance*, the navigator, with his crew, should come safely home!

PART X

Extracts and Forms from Navigation Publications

SUN

OCT.–MAR. App. Alt.	Lower Limb	Upper Limb	APR.–SEPT. App. Alt.	Lower Limb	Upper Limb
9 34	+10.8	−21.5	9 39	+10.6	−21.2
9 45	+10.9	−21.4	9 51	+10.7	−21.1
9 56	+11.0	−21.3	10 03	+10.8	−21.0
10 08	+11.1	−21.2	10 15	+10.9	−20.9
10 21	+11.2	−21.1	10 27	+11.0	−20.8
10 34	+11.3	−21.0	10 40	+11.1	−20.7
10 47	+11.4	−20.9	10 54	+11.2	−20.6
11 01	+11.5	−20.8	11 08	+11.3	−20.5
11 15	+11.6	−20.7	11 23	+11.4	−20.4
11 30	+11.7	−20.6	11 38	+11.5	−20.3
11 46	+11.8	−20.5	11 54	+11.6	−20.2
12 02	+11.9	−20.4	12 10	+11.7	−20.1
12 19	+12.0	−20.3	12 28	+11.8	−20.0
12 37	+12.1	−20.2	12 46	+11.9	−19.9
12 55	+12.2	−20.1	13 05	+12.0	−19.8
13 14	+12.3	−20.0	13 24	+12.1	−19.7
13 35	+12.4	−19.9	13 45	+12.2	−19.6
13 56	+12.5	−19.8	14 07	+12.3	−19.5
14 18	+12.6	−19.7	14 30	+12.4	−19.4
14 42	+12.7	−19.6	14 54	+12.5	−19.3
15 06	+12.8	−19.5	15 19	+12.6	−19.2
15 32	+12.9	−19.4	15 46	+12.7	−19.1
15 59	+13.0	−19.3	16 14	+12.8	−19.0
16 28	+13.1	−19.2	16 44	+12.9	−18.9
16 59	+13.2	−19.1	17 15	+13.0	−18.8
17 32	+13.3	−19.0	17 48	+13.1	−18.7
18 06	+13.4	−18.9	18 24	+13.2	−18.6
18 42	+13.5	−18.8	19 01	+13.3	−18.5
19 21	+13.6	−18.7	19 42	+13.4	−18.4
20 03	+13.7	−18.6	20 25	+13.5	−18.3
20 48	+13.8	−18.5	21 11	+13.6	−18.2
21 35	+13.9	−18.4	22 00	+13.7	−18.1
22 26	+14.0	−18.3	22 54	+13.8	−18.0
23 22	+14.1	−18.2	23 51	+13.9	−17.9
24 21	+14.2	−18.1	24 53	+14.0	−17.8
25 26	+14.3	−18.0	26 00	+14.1	−17.7
26 36	+14.4	−17.9	27 13	+14.2	−17.6
27 52	+14.5	−17.8	28 33	+14.3	−17.5
29 15	+14.6	−17.7	30 00	+14.4	−17.4
30 46	+14.7	−17.6	31 35	+14.5	−17.3
32 26	+14.8	−17.5	33 20	+14.6	−17.2
34 17	+14.9	−17.4	35 17	+14.7	−17.1
36 20	+15.0	−17.3	37 26	+14.8	−17.0
38 36	+15.1	−17.2	39 50	+14.9	−16.9
41 08	+15.2	−17.1	42 31	+15.0	−16.8
43 59	+15.3	−17.0	45 31	+15.1	−16.7
47 10	+15.4	−16.9	48 55	+15.2	−16.6
50 46	+15.5	−16.8	52 44	+15.3	−16.5
54 49	+15.6	−16.7	57 02	+15.4	−16.4
59 23	+15.7	−16.6	61 51	+15.5	−16.3
64 30	+15.8	−16.5	67 17	+15.6	−16.2
70 12	+15.9	−16.4	73 16	+15.7	−16.1
76 26	+16.0	−16.3	79 43	+15.8	−16.0
83 05	+16.1	−16.2	86 32	+15.9	−15.9
90 00			90 00		

STARS AND PLANETS

App. Alt.	Corrn
9 56	−5.3
10 08	−5.2
10 20	−5.1
10 33	−5.0
10 46	−4.9
11 00	−4.8
11 14	−4.7
11 29	−4.6
11 45	−4.5
12 01	−4.4
12 18	−4.3
12 35	−4.2
12 54	−4.1
13 13	−4.0
13 33	−3.9
13 54	−3.8
14 16	−3.7
14 40	−3.6
15 04	−3.5
15 30	−3.4
15 57	−3.3
16 26	−3.2
16 56	−3.1
17 28	−3.0
18 02	−2.9
18 38	−2.8
19 17	−2.7
19 58	−2.6
20 42	−2.5
21 28	−2.4
22 19	−2.3
23 13	−2.2
24 11	−2.1
25 14	−2.0
26 22	−1.9
27 36	−1.8
28 56	−1.7
30 24	−1.6
32 00	−1.5
33 45	−1.4
35 40	−1.3
37 48	−1.2
40 08	−1.1
42 44	−1.0
45 36	−0.9
48 47	−0.8
52 18	−0.7
56 11	−0.6
60 28	−0.5
65 08	−0.4
70 11	−0.3
75 34	−0.2
81 13	−0.1
87 03	0.0
90 00	

Additional Corrn — App. Alt.

1983

VENUS

Jan. 1–May 10
42 +0.1

May 11–June 23
47 +0.2

June 24–July 19
46 +0.3

July 20–Aug. 3
11 +0.4
41 +0.5

Aug. 4–Aug. 12
6 +0.5
20 +0.7
31

Aug. 13–Sept. 7
4 +0.6
12 +0.7
22 +0.8

Sept. 8–Sept. 16
6 +0.5
20 +0.6
31 +0.7

Sept. 17–Oct. 2
11 +0.4
41 +0.5

Oct. 3–Oct. 30
46 +0.3

Oct. 31–Dec. 17
47 +0.2

Dec. 18–Dec. 31
42 +0.1

MARS

Jan. 1–Dec. 31
60 +0.1

DIP

Ht. of Eye (m)	Corrn	Ht. of Eye (ft.)	Ht. of Eye (m)	Corrn
2.4	−2.8	8.0	1.0	−1.8
2.6	−2.9	8.6	1.5	−2.2
2.8	−3.0	9.2	2.0	−2.5
3.0	−3.1	9.8	2.5	−2.8
3.2	−3.2	10.5	3.0	−3.0
3.4	−3.3	11.2	See table ←	
3.6	−3.4	11.9	m	
3.8	−3.5	12.6	20	−7.9
4.0	−3.6	13.3	22	−8.3
4.3	−3.7	14.1	24	−8.6
4.5	−3.8	14.9	26	−9.0
4.7	−3.9	15.7	28	−9.3
5.0	−4.0	16.5	30	−9.6
5.2	−4.1	17.4	32	−10.0
5.5	−4.2	18.3	34	−10.3
5.8	−4.3	19.1	36	−10.6
6.1	−4.4	20.1	38	−10.8
6.3	−4.5	21.0	40	−11.1
6.6	−4.6	22.0	42	−11.4
6.9	−4.7	22.9	44	−11.7
7.2	−4.8	23.9	46	−11.9
7.5	−4.9	24.9	48	−12.2
7.9	−5.0	26.0	ft.	
8.2	−5.1	27.1	2	−1.4
8.5	−5.2	28.1	4	−1.9
8.8	−5.3	29.2	6	−2.4
9.2	−5.4	30.4	8	−2.7
9.5	−5.5	31.5	10	−3.1
9.9	−5.6	32.7	See table ←	
10.3	−5.7	33.9	ft.	
10.6	−5.8	35.1	70	−8.1
11.0	−5.9	36.3	75	−8.4
11.4	−6.0	37.6	80	−8.7
11.8	−6.1	38.9	85	−8.9
12.2	−6.2	40.1	90	−9.2
12.6	−6.3	41.5	95	−9.5
13.0	−6.4	42.8	100	−9.7
13.4	−6.5	44.2	105	−9.9
13.8	−6.6	45.5	110	−10.2
14.2	−6.7	46.9	115	−10.4
14.7	−6.8	48.4	120	−10.6
15.1	−6.9	49.8	125	−10.8
15.5	−7.0	51.3	130	−11.1
16.0	−7.1	52.8	135	−11.3
16.5	−7.2	54.3	140	−11.5
16.9	−7.3	55.8	145	−11.7
17.4	−7.4	57.4	150	−11.9
17.9	−7.5	58.9	155	−12.1
18.4	−7.6	60.5		
18.8	−7.7	62.1		
19.3	−7.8	63.8		
19.8	−7.9	65.4		
20.4	−8.0	67.1		
20.9	−8.1	68.8		
21.4		70.5		

App. Alt. = Apparent altitude = Sextant altitude corrected for index error and dip.
For daylight observations of Venus, see page 260.

G.M.T.	ARIES G.H.A.	VENUS −3.7 G.H.A.	Dec.	MARS +1.9 G.H.A.	Dec.	JUPITER −1.8 G.H.A.	Dec.	SATURN +0.9 G.H.A.	Dec.	STARS Name	S.H.A.	Dec.
11 00	318 53.5	161 05.2 N 2	38.3	198 19.6 N21	28.6	79 33.6 S19	46.7	111 03.1 S 8	54.8	Acamar	315 35.6	S40 22.0
01	333 56.0	176 08.5	38.0	213 20.3	28.3	94 36.0	46.7	126 05.5	54.9	Achernar	335 43.3	S57 19.0
02	348 58.5	191 11.8	37.8	228 21.1	28.0	109 38.3	46.8	141 07.8	55.0	Acrux	173 35.5	S63 00.6
03	4 00.9	206 15.1 ··	37.6	243 21.8 ··	27.7	124 40.7 ··	46.8	156 10.1 ··	55.0	Adhara	255 30.8	S28 56.7
04	19 03.4	221 18.4	37.4	258 22.6	27.4	139 43.1	46.8	171 12.4	55.1	Aldebaran	291 15.7	N16 28.6
05	34 05.9	236 21.8	37.2	273 23.3	27.1	154 45.4	46.9	186 14.7	55.2			
06	49 08.3	251 25.1 N 2	36.9	288 24.1 N21	26.8	169 47.8 S19	46.9	201 17.0 S 8	55.2	Alioth	166 40.8	N56 03.3
07	64 10.8	266 28.4	36.7	303 24.9	26.4	184 50.2	46.9	216 19.4	55.3	Alkaid	153 16.9	N49 24.1
T 08	79 13.3	281 31.7	36.5	318 25.6	26.1	199 52.5	46.9	231 21.7	55.4	Al Na'ir	28 11.7	S47 02.4
H 09	94 15.7	296 35.1 ··	36.3	333 26.4 ··	25.8	214 54.9 ··	47.0	246 24.0 ··	55.4	Alnilam	276 09.7	S 1 12.6
U 10	109 18.2	311 38.4	36.1	348 27.1	25.5	229 57.2	47.0	261 26.3	55.5	Alphard	218 18.8	S 8 35.0
R 11	124 20.6	326 41.8	35.9	3 27.9	25.2	244 59.6	47.0	276 28.6	55.5			
S 12	139 23.1	341 45.1 N 2	35.7	18 28.6 N21	24.9	260 02.0 S19	47.1	291 31.0 S 8	55.6	Alphecca	126 30.3	N26 46.4
D 13	154 25.6	356 48.5	35.5	33 29.4	24.6	275 04.3	47.1	306 33.3	55.7	Alpheratz	358 06.9	N28 59.9
A 14	169 28.0	11 51.8	35.3	48 30.1	24.2	290 06.7	47.1	321 35.6	55.7	Altair	62 30.2	N 8 49.5
Y 15	184 30.5	26 55.2 ··	35.1	63 30.9 ··	23.9	305 09.1 ··	47.2	336 37.9 ··	55.8	Ankaa	353 37.8	S42 23.6
16	199 33.0	41 58.5	34.9	78 31.7	23.6	320 11.4	47.2	351 40.2	55.9	Antares	112 54.2	S26 23.8
17	214 35.4	57 01.9	34.7	93 32.4	23.3	335 13.8	47.2	6 42.5	56.0			
18	229 37.9	72 05.3 N 2	34.5	108 33.2 N21	23.0	350 16.1 S19	47.2	21 44.9 S 8	56.0	Arcturus	146 16.6	N19 16.3
19	244 40.4	87 08.6	34.3	123 33.9	22.7	5 18.5	47.3	36 47.2	56.1	Atria	108 16.3	S69 00.2
20	259 42.8	102 12.0	34.1	138 34.7	22.4	20 20.9	47.3	51 49.5	56.1	Avior	234 28.1	S59 27.2
21	274 45.3	117 15.4 ··	33.9	153 35.5 ··	22.0	35 23.2 ··	47.3	66 51.8 ··	56.2	Bellatrix	278 56.7	N 6 20.2
22	289 47.8	132 18.8	33.7	168 36.2	21.7	50 25.6	47.4	81 54.1	56.3	Betelgeuse	271 26.2	N 7 24.4
23	304 50.2	147 22.2	33.6	183 37.0	21.4	65 27.9	47.4	96 56.4	56.3			
12 00	319 52.7	162 25.6 N 2	33.4	198 37.7 N21	21.1	80 30.3 S19	47.4	111 58.8 S 8	56.4	Canopus	264 06.7	S52 40.9
01	334 55.1	177 29.0	33.2	213 38.5	20.8	95 32.7	47.5	127 01.1	56.5	Capella	281 08.5	N45 58.8
02	349 57.6	192 32.4	33.0	228 39.3	20.4	110 35.0	47.5	142 03.4	56.5	Deneb	49 46.6	N45 13.3
03	5 00.1	207 35.8 ··	32.8	243 40.0 ··	20.1	125 37.4 ··	47.5	157 05.7 ··	56.6	Denebola	182 57.1	N14 40.1
04	20 02.5	222 39.2	32.7	258 40.8	19.8	140 39.7	47.5	172 08.0	56.7	Diphda	349 18.5	S18 04.5
05	35 05.0	237 42.6	32.5	273 41.5	19.5	155 42.1	47.6	187 10.3	56.7			
06	50 07.5	252 46.0 N 2	32.3	288 42.3 N21	19.2	170 44.5 S19	47.6	202 12.6 S 8	56.8	Dubhe	194 20.0	N61 50.7
07	65 09.9	267 49.4	32.2	303 43.1	18.9	185 46.8	47.6	217 15.0	56.8	Elnath	278 41.7	N28 35.6
08	80 12.4	282 52.8	32.0	318 43.8	18.5	200 49.1	47.7	232 17.3	56.9	Eltanin	90 56.4	N51 29.7
F 09	95 14.9	297 56.2 ··	31.8	333 44.6 ··	18.2	215 51.5 ··	47.7	247 19.6 ··	57.0	Enif	34 09.2	N 9 48.0
R 10	110 17.3	312 59.6	31.7	348 45.3	17.9	230 53.9	47.7	262 21.9	57.0	Fomalhaut	15 48.7	S29 42.5
I 11	125 19.8	328 03.1	31.5	3 46.1	17.6	245 56.2	47.8	277 24.2	57.1			
D 12	140 22.2	343 06.5 N 2	31.3	18 46.9 N21	17.2	260 58.6 S19	47.8	292 26.5 S 8	57.2	Gacrux	172 26.9	S57 01.3
A 13	155 24.7	358 09.9	31.2	33 47.6	16.9	276 00.9	47.8	307 28.8	57.2	Gienah	176 16.1	S17 26.9
Y 14	170 27.2	13 13.4	31.0	48 48.4	16.6	291 03.3	47.8	322 31.2	57.3	Hadar	149 20.7	S60 17.8
15	185 29.6	28 16.8 ··	30.9	63 49.1 ··	16.3	306 05.6 ··	47.9	337 33.5 ··	57.4	Hamal	328 26.5	N23 23.0
16	200 32.1	43 20.3	30.7	78 49.9	16.0	321 08.0	47.9	352 35.8	57.4	Kaus Aust.	84 13.8	S34 23.7
17	215 34.6	58 23.7	30.6	93 50.7	15.6	336 10.3	47.9	7 38.1	57.5			
18	230 37.0	73 27.2 N 2	30.4	108 51.4 N21	15.3	351 12.7 S19	48.0	22 40.4 S 8	57.6	Kochab	137 18.9	N74 13.7
19	245 39.5	88 30.6	30.3	123 52.2	15.0	6 15.0	48.0	37 42.7	57.6	Markab	14 00.8	N15 07.0
20	260 42.0	103 34.1	30.1	138 53.0	14.7	21 17.4	48.0	52 45.0	57.7	Menkar	314 38.9	N 4 01.6
→21	275 44.4	118 37.5 ··	30.0	153 53.7 ··	14.3	36 19.7 ··	48.1	67 47.4 ··	57.8	Menkent	148 34.7	S36 17.4
22	290 46.9	133 41.0	29.8	168 54.5	14.0	51 22.1	48.1	82 49.7	57.8	Miaplacidus	221 45.6	S69 38.9
23	305 49.4	148 44.5	29.7	183 55.2	13.7	66 24.5	48.1	97 52.0	57.9			
13 00	320 51.8	163 47.9 N 2	29.6	198 56.0 N21	13.4	81 26.8 S19	48.2	112 54.3 S 8	58.0	Mirfak	309 13.2	N49 48.0
01	335 54.3	178 51.4	29.4	213 56.8	13.1	96 29.2	48.2	127 56.6	58.0	Nunki	76 26.3	S26 19.1
02	350 56.7	193 54.9	29.3	228 57.5	12.7	111 31.5	48.2	142 58.9	58.1	Peacock	53 54.4	S56 47.4
03	5 59.2	208 58.4 ··	29.2	243 58.3 ··	12.4	126 33.9 ··	48.3	158 01.2 ··	58.2	Pollux	243 55.9	N28 04.1
04	21 01.7	224 01.8	29.0	258 59.1	12.1	141 36.2	48.3	173 03.5	58.2	Procyon	245 23.9	N 5 16.2
05	36 04.1	239 05.3	28.9	273 59.8	11.8	156 38.6	48.3	188 05.9	58.3			
06	51 06.6	254 08.8 N 2	28.8	289 00.6 N21	11.4	171 40.9 S19	48.3	203 08.2 S 8	58.4	Rasalhague	96 27.4	N12 34.4
07	66 09.1	269 12.3	28.7	304 01.4	11.1	186 43.3	48.4	218 10.5	58.4	Regulus	208 08.1	N12 03.1
S 08	81 11.5	284 15.8	28.5	319 02.1	10.8	201 45.6	48.4	233 12.8	58.5	Rigel	281 34.2	S 8 13.0
A 09	96 14.0	299 19.3 ··	28.4	334 02.9 ··	10.4	216 47.9 ··	48.4	248 15.1 ··	58.6	Rigil Kent.	140 23.2	S60 46.2
T 10	111 16.5	314 22.8	28.3	349 03.7	10.1	231 50.3	48.5	263 17.4	58.6	Sabik	102 38.6	S15 42.3
U 11	126 18.9	329 26.3	28.2	4 04.4	09.8	246 52.6	48.5	278 19.7	58.7			
R 12	141 21.4	344 29.8 N 2	28.1	19 05.2 N21	09.5	261 55.0 S19	48.5	293 22.0 S 8	58.8	Schedar	350 06.4	N56 26.6
D 13	156 23.9	359 33.3	28.0	34 06.0	09.1	276 57.3	48.6	308 24.3	58.8	Shaula	96 52.7	S37 05.7
A 14	171 26.3	14 36.9	27.8	49 06.7	08.8	291 59.7	48.6	323 26.7	58.9	Sirius	258 54.1	S16 41.4
Y 15	186 28.8	29 40.4 ··	27.7	64 07.5 ··	08.5	307 02.0 ··	48.6	338 29.0 ··	59.0	Spica	158 55.5	S11 04.4
16	201 31.2	44 43.9	27.6	79 08.2	08.2	322 04.4	48.7	353 31.3	59.0	Suhail	223 09.8	S43 21.8
17	216 33.7	59 47.4	27.5	94 09.0	07.8	337 06.7	48.7	8 33.6	59.1			
18	231 36.2	74 51.0 N 2	27.4	109 09.8 N21	07.5	352 09.1 S19	48.7	23 35.9 S 8	59.2	Vega	80 54.1	N38 46.2
19	246 38.6	89 54.5	27.3	124 10.6	07.2	7 11.4	48.8	38 38.2	59.2	Zuben'ubi	137 30.8	S15 58.4
20	261 41.1	104 58.0	27.2	139 11.3	06.8	22 13.8	48.8	53 40.5	59.3		S.H.A.	Mer. Pass.
21	276 43.6	120 01.6 ··	27.1	154 12.1 ··	06.5	37 16.1 ··	48.8	68 42.8 ··	59.4	Venus	202 32.9	13 07
22	291 46.0	135 05.1	27.0	169 12.9	06.2	52 18.4	48.9	83 45.1	59.4	Mars	238 45.1	10 45
23	306 48.5	150 08.7	26.9	184 13.6	05.9	67 20.8	48.9	98 47.5	59.5	Jupiter	120 37.6	18 35
Mer. Pass. 2 40.0		v 3.4	d 0.2	v 0.8	d 0.3	v 2.4	d 0.0	v 2.3	d 0.1	Saturn	152 06.1	16 30

G.M.T.	SUN G.H.A.	SUN Dec.	MOON G.H.A.	v	MOON Dec.	d	H.P.
11 00	178 40.3	N15 30.4	146 59.6	8.9	N 9 01.6	14.5	60.6
01	193 40.4	29.6	161 27.5	8.9	8 47.1	14.5	60.6
02	208 40.5	28.9	175 55.4	9.0	8 32.6	14.5	60.6
03	223 40.6	·· 28.2	190 23.4	9.0	8 18.1	14.6	60.5
04	238 40.7	27.4	204 51.4	9.2	8 03.5	14.6	60.5
05	253 40.8	26.7	219 19.6	9.2	7 48.9	14.7	60.5
06	268 40.9	N15 25.9	233 47.8	9.3	N 7 34.2	14.7	60.5
07	283 41.0	25.2	248 16.1	9.3	7 19.5	14.7	60.4
T 08	298 41.1	24.5	262 44.4	9.5	7 04.8	14.7	60.4
H 09	313 41.2	·· 23.7	277 12.9	9.5	6 50.1	14.7	60.4
U 10	328 41.3	23.0	291 41.4	9.5	6 35.4	14.8	60.3
R 11	343 41.4	22.3	306 09.9	9.7	6 20.6	14.8	60.3
S 12	358 41.5	N15 21.5	320 38.6	9.7	N 6 05.8	14.8	60.3
D 13	13 41.6	20.8	335 07.3	9.8	5 51.0	14.9	60.2
A 14	28 41.7	20.0	349 36.1	9.8	5 36.1	14.9	60.2
Y 15	43 41.8	·· 19.3	4 04.9	10.0	5 21.3	14.9	60.2
16	58 41.9	18.6	18 33.9	9.9	5 06.4	14.9	60.1
17	73 42.0	17.8	33 02.8	10.1	4 51.5	14.8	60.1
18	88 42.1	N15 17.1	47 31.9	10.1	N 4 36.7	14.9	60.1
19	103 42.2	16.3	62 01.0	10.2	4 21.8	14.9	60.0
20	118 42.3	15.6	76 30.2	10.2	4 06.9	14.9	60.0
21	133 42.4	·· 14.9	90 59.4	10.3	3 52.0	15.0	60.0
22	148 42.5	14.1	105 28.7	10.3	3 37.0	14.9	59.9
23	163 42.6	13.4	119 58.0	10.4	3 22.1	14.9	59.9
12 00	178 42.7	N15 12.6	134 27.4	10.5	N 3 07.2	14.9	59.9
01	193 42.8	11.9	148 56.9	10.5	2 52.3	14.9	59.8
02	208 42.9	11.1	163 26.4	10.6	2 37.4	14.9	59.8
03	223 43.0	·· 10.4	177 56.0	10.6	2 22.5	14.9	59.8
04	238 43.1	09.7	192 25.6	10.7	2 07.6	14.9	59.7
05	253 43.2	08.9	206 55.3	10.7	1 52.7	14.9	59.7
06	268 43.3	N15 08.2	221 25.0	10.8	N 1 37.8	14.9	59.7
07	283 43.4	07.4	235 54.8	10.8	1 23.0	14.9	59.6
08	298 43.5	06.7	250 24.6	10.9	1 08.1	14.9	59.6
F 09	313 43.6	·· 05.9	264 54.5	11.0	0 53.2	14.8	59.5
R 10	328 43.7	05.2	279 24.5	10.9	0 38.4	14.8	59.5
I 11	343 43.8	04.4	293 54.4	11.0	0 23.6	14.8	59.5
D 12	358 43.9	N15 03.7	308 24.4	11.1	N 0 08.8	14.8	59.4
A 13	13 44.0	02.9	322 54.5	11.1	S 0 06.0	14.8	59.4
Y 14	28 44.1	02.2	337 24.6	11.1	0 20.8	14.7	59.4
15	43 44.2	·· 01.4	351 54.7	11.2	0 35.5	14.7	59.3
16	58 44.3	15 00.7	6 24.9	11.3	0 50.2	14.7	59.3
17	73 44.4	14 59.9	20 55.2	11.2	1 04.9	14.7	59.3
18	88 44.5	N14 59.2	35 25.4	11.3	S 1 19.6	14.6	59.2
19	103 44.6	58.4	49 55.7	11.4	1 34.2	14.6	59.2
20	118 44.7	57.7	64 26.1	11.3	1 48.8	14.6	59.1
21	133 44.8	·· 56.9	78 56.4	11.4	2 03.4	14.6	59.1
22	148 44.9	56.2	93 26.8	11.5	2 18.0	14.5	59.1
23	163 45.0	55.4	107 57.3	11.4	2 32.5	14.5	59.0
13 00	178 45.1	N14 54.7	122 27.7	11.5	S 2 47.0	14.5	59.0
01	193 45.3	53.9	136 58.2	11.6	3 01.5	14.4	58.9
02	208 45.4	53.2	151 28.8	11.5	3 15.9	14.4	58.9
03	223 45.5	·· 52.4	165 59.3	11.6	3 30.3	14.3	58.9
04	238 45.6	51.7	180 29.9	11.6	3 44.6	14.3	58.8
05	253 45.7	50.9	195 00.5	11.7	3 58.9	14.3	58.8
06	268 45.8	N14 50.1	209 31.2	11.7	S 4 13.2	14.2	58.7
07	283 45.9	49.4	224 01.9	11.6	4 27.4	14.2	58.7
S 08	298 46.0	48.6	238 32.5	11.8	4 41.6	14.1	58.7
A 09	313 46.1	·· 47.9	253 03.3	11.7	4 55.7	14.1	58.6
T 10	328 46.2	47.1	267 34.0	11.8	5 09.8	14.0	58.6
U 11	343 46.3	46.4	282 04.8	11.7	5 23.9	14.0	58.6
R 12	358 46.4	N14 45.6	296 35.5	11.8	S 5 37.9	13.9	58.5
D 13	13 46.5	44.8	311 06.3	11.8	5 51.8	13.9	58.5
A 14	28 46.7	44.1	325 37.1	11.9	6 05.7	13.9	58.4
Y 15	43 46.8	·· 43.3	340 08.0	11.8	6 19.6	13.8	58.4
16	58 46.9	42.6	354 38.8	11.9	6 33.4	13.8	58.4
17	73 47.0	41.8	9 09.7	11.9	6 47.2	13.7	58.3
18	88 47.1	N14 41.0	23 40.6	11.8	S 7 00.9	13.6	58.3
19	103 47.2	40.3	38 11.4	11.9	7 14.5	13.6	58.2
20	118 47.3	39.5	52 42.3	12.0	7 28.1	13.5	58.2
21	133 47.4	·· 38.8	67 13.3	11.9	7 41.6	13.5	58.2
22	148 47.5	38.0	81 44.2	11.9	7 55.1	13.4	58.1
23	163 47.7	37.2	96 15.1	11.9	8 08.5	13.4	58.1
	S.D. 15.8	d 0.7	S.D. 16.4		16.2		15.9

Lat.	Twilight Naut.	Twilight Civil	Sunrise	Moonrise 11	Moonrise 12	Moonrise 13	Moonrise 14
N 72	////	////	01 57	06 56	09 13	11 23	13 37
N 70	////	////	02 37	07 07	09 14	11 15	13 17
68	////	01 09	03 04	07 16	09 14	11 08	13 01
66	////	02 01	03 24	07 23	09 15	11 02	12 48
64	////	02 31	03 40	07 29	09 16	10 58	12 37
62	01 02	02 54	03 54	07 35	09 16	10 53	12 29
60	01 48	03 12	04 05	07 39	09 16	10 50	12 21
N 58	02 16	03 27	04 15	07 43	09 17	10 47	12 14
56	02 37	03 39	04 24	07 47	09 17	10 44	12 08
54	02 54	03 50	04 31	07 50	09 18	10 42	12 03
52	03 08	03 59	04 38	07 53	09 18	10 39	11 58
50	03 20	04 07	04 44	07 56	09 18	10 37	11 54
45	03 45	04 25	04 57	08 02	09 19	10 33	11 45
N 40	04 03	04 39	05 08	08 06	09 19	10 29	11 37
35	04 17	04 50	05 17	08 10	09 20	10 26	11 31
30	04 30	05 00	05 25	08 14	09 20	10 23	11 25
20	04 49	05 16	05 39	08 20	09 21	10 19	11 15
N 10	05 03	05 29	05 51	08 26	09 21	10 14	11 07
0	05 15	05 40	06 02	08 31	09 22	10 11	10 59
S 10	05 26	05 51	06 13	08 36	09 22	10 07	10 51
20	05 35	06 01	06 24	08 41	09 23	10 03	10 42
30	05 44	06 12	06 37	08 47	09 24	09 58	10 33
35	05 48	06 18	06 44	08 51	09 24	09 56	10 28
40	05 53	06 25	06 53	08 55	09 25	09 53	10 22
45	05 57	06 32	07 03	09 00	09 25	09 50	10 14
S 50	06 02	06 40	07 15	09 05	09 26	09 46	10 06
52	06 04	06 44	07 20	09 08	09 26	09 44	10 02
54	06 06	06 48	07 26	09 10	09 27	09 42	09 58
56	06 09	06 53	07 33	09 13	09 27	09 40	09 53
58	06 11	06 58	07 40	09 17	09 27	09 37	09 48
S 60	06 14	07 03	07 49	09 21	09 28	09 35	09 42

Lat.	Sunset	Twilight Civil	Twilight Naut.	Moonset 11	Moonset 12	Moonset 13	Moonset 14
N 72	22 07	////	////	21 41	21 17	20 51	20 19
N 70	21 29	////	////	21 36	21 20	21 02	20 42
68	21 03	22 51	////	21 33	21 22	21 11	20 59
66	20 43	22 05	////	21 29	21 24	21 19	21 13
64	20 27	21 35	////	21 26	21 26	21 25	21 25
62	20 14	21 13	23 00	21 24	21 28	21 31	21 35
60	20 03	20 56	22 18	21 22	21 29	21 36	21 44
N 58	19 54	20 42	21 51	21 20	21 30	21 40	21 52
56	19 45	20 30	21 30	21 18	21 31	21 44	21 59
54	19 38	20 19	21 14	21 16	21 32	21 48	22 05
52	19 31	20 10	21 00	21 15	21 33	21 51	22 10
50	19 25	20 02	20 48	21 14	21 34	21 54	22 15
45	19 12	19 44	20 24	21 11	21 36	22 00	22 26
N 40	19 02	19 31	20 06	21 08	21 37	22 06	22 35
35	18 53	19 20	19 52	21 06	21 39	22 10	22 43
30	18 45	19 10	19 40	21 04	21 40	22 14	22 50
20	18 31	18 54	19 21	21 01	21 42	22 21	23 01
N 10	18 19	18 41	19 07	20 58	21 43	22 28	23 12
0	18 08	18 30	18 55	20 55	21 45	22 34	23 22
S 10	17 58	18 19	18 45	20 52	21 47	22 39	23 31
20	17 46	18 09	18 35	20 49	21 48	22 46	23 42
30	17 33	17 58	18 27	20 46	21 50	22 53	23 54
35	17 26	17 52	18 22	20 44	21 52	22 57	24 01
40	17 18	17 46	18 18	20 41	21 53	23 02	24 09
45	17 08	17 39	18 13	20 39	21 54	23 07	24 19
S 50	16 56	17 30	18 09	20 35	21 56	23 14	24 30
52	16 51	17 27	18 07	20 34	21 57	23 17	24 35
54	16 45	17 23	18 05	20 32	21 58	23 21	24 41
56	16 38	17 18	18 02	20 30	21 59	23 24	24 48
58	16 31	17 13	18 00	20 28	22 00	23 28	24 55
S 60	16 22	17 08	17 57	20 26	22 01	23 33	25 03

Day	SUN Eqn. of Time 00h	SUN Eqn. of Time 12h	SUN Mer. Pass.	MOON Mer. Pass. Upper	MOON Mer. Pass. Lower	Age	Phase
	m s	m s	h m	h m	h m	d	
11	05 43	05 14	12 05	14 43	02 17	03	
12	05 10	05 05	12 05	15 33	03 09	04	◐
13	05 00	04 54	12 05	16 22	03 58	05	

CONVERSION OF ARC TO TIME

°	h m	°	h m	°	h m	°	h m	°	h m	°	h m	′	0 00 m s	0′·25 m s	0′·50 m s	0′·75 m s
0	0 00	60	4 00	120	8 00	180	12 00	240	16 00	300	20 00	0	0 00	0 01	0 02	0 03
1	0 04	61	4 04	121	8 04	181	12 04	241	16 04	301	20 04	1	0 04	0 05	0 06	0 07
2	0 08	62	4 08	122	8 08	182	12 08	242	16 08	302	20 08	2	0 08	0 09	0 10	0 11
3	0 12	63	4 12	123	8 12	183	12 12	243	16 12	303	20 12	3	0 12	0 13	0 14	0 15
4	0 16	64	4 16	124	8 16	184	12 16	244	16 16	304	20 16	4	0 16	0 17	0 18	0 19
5	0 20	65	4 20	125	8 20	185	12 20	245	16 20	305	20 20	5	0 20	0 21	0 22	0 23
6	0 24	66	4 24	126	8 24	186	12 24	246	16 24	306	20 24	6	0 24	0 25	0 26	0 27
7	0 28	67	4 28	127	8 28	187	12 28	247	16 28	307	20 28	7	0 28	0 29	0 30	0 31
8	0 32	68	4 32	128	8 32	188	12 32	248	16 32	308	20 32	8	0 32	0 33	0 34	0 35
9	0 36	69	4 36	129	8 36	189	12 36	249	16 36	309	20 36	9	0 36	0 37	0 38	0 39
10	0 40	70	4 40	130	8 40	190	12 40	250	16 40	310	20 40	10	0 40	0 41	0 42	0 43
11	0 44	71	4 44	131	8 44	191	12 44	251	16 44	311	20 44	11	0 44	0 45	0 46	0 47
12	0 48	72	4 48	132	8 48	192	12 48	252	16 48	312	20 48	12	0 48	0 49	0 50	0 51
13	0 52	73	4 52	133	8 52	193	12 52	253	16 52	313	20 52	13	0 52	0 53	0 54	0 55
14	0 56	74	4 56	134	8 56	194	12 56	254	16 56	314	20 56	14	0 56	0 57	0 58	0 59
15	1 00	75	5 00	135	9 00	195	13 00	255	17 00	315	21 00	15	1 00	1 01	1 02	1 03
16	1 04	76	5 04	136	9 04	196	13 04	256	17 04	316	21 04	16	1 04	1 05	1 06	1 07
17	1 08	77	5 08	137	9 08	197	13 08	257	17 08	317	21 08	17	1 08	1 09	1 10	1 11
18	1 12	78	5 12	138	9 12	198	13 12	258	17 12	318	21 12	18	1 12	1 13	1 14	1 15
19	1 16	79	5 16	139	9 16	199	13 16	259	17 16	319	21 16	19	1 16	1 17	1 18	1 19
20	1 20	80	5 20	140	9 20	200	13 20	260	17 20	320	21 20	20	1 20	1 21	1 22	1 23
21	1 24	81	5 24	141	9 24	201	13 24	261	17 24	321	21 24	21	1 24	1 25	1 26	1 27
22	1 28	82	5 28	142	9 28	202	13 28	262	17 28	322	21 28	22	1 28	1 29	1 30	1 31
23	1 32	83	5 32	143	9 32	203	13 32	263	17 32	323	21 32	23	1 32	1 33	1 34	1 35
24	1 36	84	5 36	144	9 36	204	13 36	264	17 36	324	21 36	24	1 36	1 37	1 38	1 39
25	1 40	85	5 40	145	9 40	205	13 40	265	17 40	325	21 40	25	1 40	1 41	1 42	1 43
26	1 44	86	5 44	146	9 44	206	13 44	266	17 44	326	21 44	26	1 44	1 45	1 46	1 47
27	1 48	87	5 48	147	9 48	207	13 48	267	17 48	327	21 48	27	1 48	1 49	1 50	1 51
28	1 52	88	5 52	148	9 52	208	13 52	268	17 52	328	21 52	28	1 52	1 53	1 54	1 55
29	1 56	89	5 56	149	9 56	209	13 56	269	17 56	329	21 56	29	1 56	1 57	1 58	1 59
30	2 00	90	6 00	150	10 00	210	14 00	270	18 00	330	22 00	30	2 00	2 01	2 02	2 03
31	2 04	91	6 04	151	10 04	211	14 04	271	18 04	331	22 04	31	2 04	2 05	2 06	2 07
32	2 08	92	6 08	152	10 08	212	14 08	272	18 08	332	22 08	32	2 08	2 09	2 10	2 11
33	2 12	93	6 12	153	10 12	213	14 12	273	18 12	333	22 12	33	2 12	2 13	2 14	2 15
34	2 16	94	6 16	154	10 16	214	14 16	274	18 16	334	22 16	34	2 16	2 17	2 18	2 19
35	2 20	95	6 20	155	10 20	215	14 20	275	18 20	335	22 20	35	2 20	2 21	2 22	2 23
36	2 24	96	6 24	156	10 24	216	14 24	276	18 24	336	22 24	36	2 24	2 25	2 26	2 27
37	2 28	97	6 28	157	10 28	217	14 28	277	18 28	337	22 28	37	2 28	2 29	2 30	2 31
38	2 32	98	6 32	158	10 32	218	14 32	278	18 32	338	22 32	38	2 32	2 33	2 34	2 35
39	2 36	99	6 36	159	10 36	219	14 36	279	18 36	339	22 36	39	2 36	2 37	2 38	2 39
40	2 40	100	6 40	160	10 40	220	14 40	280	18 40	340	22 40	40	2 40	2 41	2 42	2 43
41	2 44	101	6 44	161	10 44	221	14 44	281	18 44	341	22 44	41	2 44	2 45	2 46	2 47
42	2 48	102	6 48	162	10 48	222	14 48	282	18 48	342	22 48	42	2 48	2 49	2 50	2 51
43	2 52	103	6 52	163	10 52	223	14 52	283	18 52	343	22 52	43	2 52	2 53	2 54	2 55
44	2 56	104	6 56	164	10 56	224	14 56	284	18 56	344	22 56	44	2 56	2 57	2 58	2 59
45	3 00	105	7 00	165	11 00	225	15 00	285	19 00	345	23 00	45	3 00	3 01	3 02	3 03
46	3 04	106	7 04	166	11 04	226	15 04	286	19 04	346	23 04	46	3 04	3 05	3 06	3 07
47	3 08	107	7 08	167	11 08	227	15 08	287	19 08	347	23 08	47	3 08	3 09	3 10	3 11
48	3 12	108	7 12	168	11 12	228	15 12	288	19 12	348	23 12	·48	3 12	3 13	3 14	3 15
49	3 16	109	7 16	169	11 16	229	15 16	289	19 16	349	23 16	49	3 16	3 17	3 18	3 19
50	3 20	110	7 20	170	11 20	230	15 20	290	19 20	350	23 20	50	3 20	3 21	3 22	3 23
51	3 24	111	7 24	171	11 24	231	15 24	291	19 24	351	23 24	51	3 24	3 25	3 26	3 27
52	3 28	112	7 28	172	11 28	232	15 28	292	19 28	352	23 28	52	3 28	3 29	3 30	3 31
53	3 32	113	7 32	173	11 32	233	15 32	293	19 32	353	23 32	53	3 32	3 33	3 34	3 35
54	3 36	114	7 36	174	11 36	234	15 36	294	19 36	354	23 36	54	3 36	3 37	3 38	3 39
55	3 40	115	7 40	175	11 40	235	15 40	295	19 40	355	23 40	55	3 40	3 41	3 42	3 43
56	3 44	116	7 44	176	11 44	236	15 44	296	19 44	356	23 44	56	3 44	3 45	3 46	3 47
57	3 48	117	7 48	177	11 48	237	15 48	297	19 48	357	23 48	57	3 48	3 49	3 50	3 51
58	3 52	118	7 52	178	11 52	238	15 52	298	19 52	358	23 52	58	3 52	3 53	3 54	3 55
59	3 56	119	7 56	179	11 56	239	15 56	299	19 56	359	23 56	59	3 56	3 57	3 58	3 59

The above table is for converting expressions in arc to their equivalent in time ; its main use in this Almanac is for the conversion of longitude for application to L.M.T. (*added* if *west*, *subtracted* if *east*) to give G.M.T. or vice versa, particularly in the case of sunrise, sunset, etc.

761-083 O - 66 - 19

12ᵐ	SUN PLANETS	ARIES	MOON	v or d Corrⁿ	v or d Corrⁿ	v or d Corrⁿ
s	° ′	° ′	° ′	′ ′	′ ′	′ ′
00	3 00·0	3 00·5	2 51·8	0·0 0·0	6·0 1·3	12·0 2·5
01	3 00·3	3 00·7	2 52·0	0·1 0·0	6·1 1·3	12·1 2·5
02	3 00·5	3 01·0	2 52·3	0·2 0·0	6·2 1·3	12·2 2·5
03	3 00·8	3 01·2	2 52·5	0·3 0·1	6·3 1·3	12·3 2·6
04	3 01·0	3 01·5	2 52·8	0·4 0·1	6·4 1·3	12·4 2·6
05	3 01·3	3 01·7	2 53·0	0·5 0·1	6·5 1·4	12·5 2·6
06	3 01·5	3 02·0	2 53·2	0·6 0·1	6·6 1·4	12·6 2·6
07	3 01·8	3 02·2	2 53·5	0·7 0·1	6·7 1·4	12·7 2·6
08	3 02·0	3 02·5	2 53·7	0·8 0·2	6·8 1·4	12·8 2·7
09	3 02·3	3 02·7	2 53·9	0·9 0·2	6·9 1·4	12·9 2·7
10	3 02·5	3 03·0	2 54·2	1·0 0·2	7·0 1·5	13·0 2·7
11	3 02·8	3 03·3	2 54·4	1·1 0·2	7·1 1·5	13·1 2·7
12	3 03·0	3 03·5	2 54·7	1·2 0·3	7·2 1·5	13·2 2·8
13	3 03·3	3 03·8	2 54·9	1·3 0·3	7·3 1·5	13·3 2·8
14	3 03·5	3 04·0	2 55·1	1·4 0·3	7·4 1·5	13·4 2·8
15	3 03·8	3 04·3	2 55·4	1·5 0·3	7·5 1·6	13·5 2·8
16	3 04·0	3 04·5	2 55·6	1·6 0·3	7·6 1·6	13·6 2·8
17	3 04·3	3 04·8	2 55·9	1·7 0·4	7·7 1·6	13·7 2·9
18	3 04·5	3 05·0	2 56·1	1·8 0·4	7·8 1·6	13·8 2·9
19	3 04·8	3 05·3	2 56·3	1·9 0·4	7·9 1·6	13·9 2·9
20	3 05·0	3 05·5	2 56·6	2·0 0·4	8·0 1·7	14·0 2·9
21	3 05·3	3 05·8	2 56·8	2·1 0·4	8·1 1·7	14·1 2·9
22	3 05·5	3 06·0	2 57·0	2·2 0·5	8·2 1·7	14·2 3·0
23	3 05·8	3 06·3	2 57·3	2·3 0·5	8·3 1·7	14·3 3·0
24	3 06·0	3 06·5	2 57·5	2·4 0·5	8·4 1·8	14·4 3·0
25	3 06·3	3 06·8	2 57·8	2·5 0·5	8·5 1·8	14·5 3·0
26	3 06·5	3 07·0	2 58·0	2·6 0·5	8·6 1·8	14·6 3·0
27	3 06·8	3 07·3	2 58·2	2·7 0·6	8·7 1·8	14·7 3·1
28	3 07·0	3 07·5	2 58·5	2·8 0·6	8·8 1·8	14·8 3·1
29	3 07·3	3 07·8	2 58·7	2·9 0·6	8·9 1·9	14·9 3·1
30	3 07·5	3 08·0	2 59·0	3·0 0·6	9·0 1·9	15·0 3·1
31	3 07·8	3 08·3	2 59·2	3·1 0·6	9·1 1·9	15·1 3·1
32	3 08·0	3 08·5	2 59·4	3·2 0·7	9·2 1·9	15·2 3·2
33	3 08·3	3 08·8	2 59·7	3·3 0·7	9·3 1·9	15·3 3·2
34	3 08·5	3 09·0	2 59·9	3·4 0·7	9·4 2·0	15·4 3·2
35	3 08·8	3 09·3	3 00·2	3·5 0·7	9·5 2·0	15·5 3·2
36	3 09·0	3 09·5	3 00·4	3·6 0·8	9·6 2·0	15·6 3·3
37	3 09·3	3 09·8	3 00·6	3·7 0·8	9·7 2·0	15·7 3·3
38	3 09·5	3 10·0	3 00·9	3·8 0·8	9·8 2·0	15·8 3·3
39	3 09·8	3 10·3	3 01·1	3·9 0·8	9·9 2·1	15·9 3·3
40	3 10·0	3 10·5	3 01·3	4·0 0·8	10·0 2·1	16·0 3·3
41	3 10·3	3 10·8	3 01·6	4·1 0·9	10·1 2·1	16·1 3·4
42	3 10·5	3 11·0	3 01·8	4·2 0·9	10·2 2·1	16·2 3·4
43	3 10·8	3 11·3	3 02·1	4·3 0·9	10·3 2·1	16·3 3·4
44	3 11·0	3 11·5	3 02·3	4·4 0·9	10·4 2·2	16·4 3·4
45	3 11·3	3 11·8	3 02·5	4·5 0·9	10·5 2·2	16·5 3·4
46	3 11·5	3 12·0	3 02·8	4·6 1·0	10·6 2·2	16·6 3·5
47	3 11·8	3 12·3	3 03·0	4·7 1·0	10·7 2·2	16·7 3·5
48	3 12·0	3 12·5	3 03·3	4·8 1·0	10·8 2·3	16·8 3·5
49	3 12·3	3 12·8	3 03·5	4·9 1·0	10·9 2·3	16·9 3·5
50	3 12·5	3 13·0	3 03·7	5·0 1·0	11·0 2·3	17·0 3·5
51	3 12·8	3 13·3	3 04·0	5·1 1·1	11·1 2·3	17·1 3·6
52	3 13·0	3 13·5	3 04·2	5·2 1·1	11·2 2·3	17·2 3·6
53	3 13·3	3 13·8	3 04·4	5·3 1·1	11·3 2·4	17·3 3·6
54	3 13·5	3 14·0	3 04·7	5·4 1·1	11·4 2·4	17·4 3·6
55	3 13·8	3 14·3	3 04·9	5·5 1·1	11·5 2·4	17·5 3·6
56	3 14·0	3 14·5	3 05·2	5·6 1·2	11·6 2·4	17·6 3·7
57	3 14·3	3 14·8	3 05·4	5·7 1·2	11·7 2·4	17·7 3·7
58	3 14·5	3 15·0	3 05·6	5·8 1·2	11·8 2·5	17·8 3·7
59	3 14·8	3 15·3	3 05·9	5·9 1·2	11·9 2·5	17·9 3·7
60	3 15·0	3 15·5	3 06·1	6·0 1·3	12·0 2·5	18·0 3·8

13ᵐ	SUN PLANETS	ARIES	MOON	v or d Corrⁿ	v or d Corrⁿ	v or d Corrⁿ
s	° ′	° ′	° ′	′ ′	′ ′	′ ′
00	3 15·0	3 15·5	3 06·1	0·0 0·0	6·0 1·4	12·0 2·7
01	3 15·3	3 15·8	3 06·4	0·1 0·0	6·1 1·4	12·1 2·7
02	3 15·5	3 16·0	3 06·6	0·2 0·0	6·2 1·4	12·2 2·7
03	3 15·8	3 16·3	3 06·8	0·3 0·1	6·3 1·4	12·3 2·8
04	3 16·0	3 16·5	3 07·1	0·4 0·1	6·4 1·4	12·4 2·8
05	3 16·3	3 16·8	3 07·3	0·5 0·1	6·5 1·5	12·5 2·8
06	3 16·5	3 17·0	3 07·5	0·6 0·1	6·6 1·5	12·6 2·8
07	3 16·8	3 17·3	3 07·8	0·7 0·2	6·7 1·5	12·7 2·9
08	3 17·0	3 17·5	3 08·0	0·8 0·2	6·8 1·5	12·8 2·9
09	3 17·3	3 17·8	3 08·3	0·9 0·2	6·9 1·6	12·9 2·9
10	3 17·5	3 18·0	3 08·5	1·0 0·2	7·0 1·6	13·0 2·9
11	3 17·8	3 18·3	3 08·7	1·1 0·2	7·1 1·6	13·1 2·9
12	3 18·0	3 18·5	3 09·0	1·2 0·3	7·2 1·6	13·2 3·0
13	3 18·3	3 18·8	3 09·2	1·3 0·3	7·3 1·6	13·3 3·0
14	3 18·5	3 19·0	3 09·5	1·4 0·3	7·4 1·7	13·4 3·0
15	3 18·8	3 19·3	3 09·7	1·5 0·3	7·5 1·7	13·5 3·0
16	3 19·0	3 19·5	3 09·9	1·6 0·4	7·6 1·7	13·6 3·1
17	3 19·3	3 19·8	3 10·2	1·7 0·4	7·7 1·7	13·7 3·1
18	3 19·5	3 20·0	3 10·4	1·8 0·4	7·8 1·8	13·8 3·1
19	3 19·8	3 20·3	3 10·7	1·9 0·4	7·9 1·8	13·9 3·1
20	3 20·0	3 20·5	3 10·9	2·0 0·5	8·0 1·8	14·0 3·2
21	3 20·3	3 20·8	3 11·1	2·1 0·5	8·1 1·8	14·1 3·2
22	3 20·5	3 21·0	3 11·4	2·2 0·5	8·2 1·8	14·2 3·2
23	3 20·8	3 21·3	3 11·6	2·3 0·5	8·3 1·9	14·3 3·2
24	3 21·0	3 21·6	3 11·8	2·4 0·5	8·4 1·9	14·4 3·2
25	3 21·3	3 21·8	3 12·1	2·5 0·6	8·5 1·9	14·5 3·3
26	3 21·5	3 22·1	3 12·3	2·6 0·6	8·6 1·9	14·6 3·3
27	3 21·8	3 22·3	3 12·6	2·7 0·6	8·7 2·0	14·7 3·3
28	3 22·0	3 22·6	3 12·8	2·8 0·6	8·8 2·0	14·8 3·3
29	3 22·3	3 22·8	3 13·0	2·9 0·7	8·9 2·0	14·9 3·4
30	3 22·5	3 23·1	3 13·3	3·0 0·7	9·0 2·0	15·0 3·4
31	3 22·8	3 23·3	3 13·5	3·1 0·7	9·1 2·0	15·1 3·4
32	3 23·0	3 23·6	3 13·8	3·2 0·7	9·2 2·1	15·2 3·4
33	3 23·3	3 23·8	3 14·0	3·3 0·7	9·3 2·1	15·3 3·4
34	3 23·5	3 24·1	3 14·2	3·4 0·8	9·4 2·1	15·4 3·5
35	3 23·8	3 24·3	3 14·5	3·5 0·8	9·5 2·1	15·5 3·5
36	3 24·0	3 24·6	3 14·7	3·6 0·8	9·6 2·2	15·6 3·5
37	3 24·3	3 24·8	3 14·9	3·7 0·8	9·7 2·2	15·7 3·5
38	3 24·5	3 25·1	3 15·2	3·8 0·9	9·8 2·2	15·8 3·6
39	3 24·8	3 25·3	3 15·4	3·9 0·9	9·9 2·2	15·9 3·6
40	3 25·0	3 25·6	3 15·7	4·0 0·9	10·0 2·3	16·0 3·6
41	3 25·3	3 25·8	3 15·9	4·1 0·9	10·1 2·3	16·1 3·6
42	3 25·5	3 26·1	3 16·1	4·2 0·9	10·2 2·3	16·2 3·6
43	3 25·8	3 26·3	3 16·4	4·3 1·0	10·3 2·3	16·3 3·7
44	3 26·0	3 26·6	3 16·6	4·4 1·0	10·4 2·3	16·4 3·7
45	3 26·3	3 26·8	3 16·9	4·5 1·0	10·5 2·4	16·5 3·7
46	3 26·5	3 27·1	3 17·1	4·6 1·0	10·6 2·4	16·6 3·7
47	3 26·8	3 27·3	3 17·3	4·7 1·1	10·7 2·4	16·7 3·8
48	3 27·0	3 27·6	3 17·6	4·8 1·1	10·8 2·4	16·8 3·8
49	3 27·3	3 27·8	3 17·8	4·9 1·1	10·9 2·5	16·9 3·8
50	3 27·5	3 28·1	3 18·0	5·0 1·1	11·0 2·5	17·0 3·8
51	3 27·8	3 28·3	3 18·3	5·1 1·1	11·1 2·5	17·1 3·8
52	3 28·0	3 28·6	3 18·5	5·2 1·2	11·2 2·5	17·2 3·9
53	3 28·3	3 28·8	3 18·8	5·3 1·2	11·3 2·5	17·3 3·9
54	3 28·5	3 29·1	3 19·0	5·4 1·2	11·4 2·6	17·4 3·9
55	3 28·8	3 29·3	3 19·2	5·5 1·2	11·5 2·6	17·5 3·9
56	3 29·0	3 29·6	3 19·5	5·6 1·3	11·6 2·6	17·6 4·0
57	3 29·3	3 29·8	3 19·7	5·7 1·3	11·7 2·6	17·7 4·0
58	3 29·5	3 30·1	3 20·0	5·8 1·3	11·8 2·7	17·8 4·0
59	3 29·8	3 30·3	3 20·2	5·9 1·3	11·9 2·7	17·9 4·0
60	3 30·0	3 30·6	3 20·4	6·0 1·4	12·0 2·7	18·0 4·1

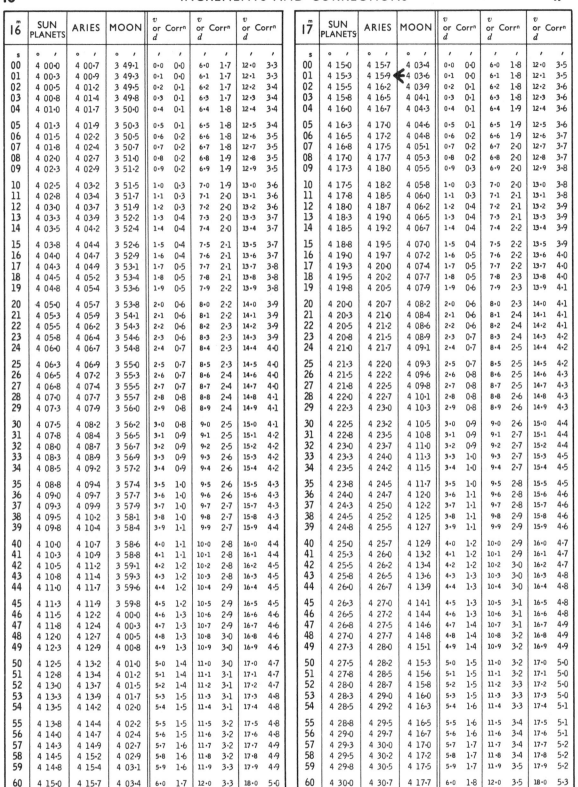

16ᵐ s	SUN PLANETS ° ′	ARIES ° ′	MOON ° ′	v or d ′	Corrⁿ ′	v or d ′	Corrⁿ ′	v or d ′	Corrⁿ ′
00	4 00·0	4 00·7	3 49·1	0·0	0·0	6·0	1·7	12·0	3·3
01	4 00·3	4 00·9	3 49·3	0·1	0·0	6·1	1·7	12·1	3·3
02	4 00·5	4 01·2	3 49·5	0·2	0·1	6·2	1·7	12·2	3·4
03	4 00·8	4 01·4	3 49·8	0·3	0·1	6·3	1·7	12·3	3·4
04	4 01·0	4 01·7	3 50·0	0·4	0·1	6·4	1·8	12·4	3·4
05	4 01·3	4 01·9	3 50·3	0·5	0·1	6·5	1·8	12·5	3·4
06	4 01·5	4 02·2	3 50·5	0·6	0·2	6·6	1·8	12·6	3·5
07	4 01·8	4 02·4	3 50·7	0·7	0·2	6·7	1·8	12·7	3·5
08	4 02·0	4 02·7	3 51·0	0·8	0·2	6·8	1·9	12·8	3·5
09	4 02·3	4 02·9	3 51·2	0·9	0·2	6·9	1·9	12·9	3·5
10	4 02·5	4 03·2	3 51·5	1·0	0·3	7·0	1·9	13·0	3·6
11	4 02·8	4 03·4	3 51·7	1·1	0·3	7·1	2·0	13·1	3·6
12	4 03·0	4 03·7	3 51·9	1·2	0·3	7·2	2·0	13·2	3·6
13	4 03·3	4 03·9	3 52·2	1·3	0·4	7·3	2·0	13·3	3·7
14	4 03·5	4 04·2	3 52·4	1·4	0·4	7·4	2·0	13·4	3·7
15	4 03·8	4 04·4	3 52·6	1·5	0·4	7·5	2·1	13·5	3·7
16	4 04·0	4 04·7	3 52·9	1·6	0·4	7·6	2·1	13·6	3·7
17	4 04·3	4 04·9	3 53·1	1·7	0·5	7·7	2·1	13·7	3·8
18	4 04·5	4 05·2	3 53·4	1·8	0·5	7·8	2·1	13·8	3·8
19	4 04·8	4 05·4	3 53·6	1·9	0·5	7·9	2·2	13·9	3·8
20	4 05·0	4 05·7	3 53·8	2·0	0·6	8·0	2·2	14·0	3·9
21	4 05·3	4 05·9	3 54·1	2·1	0·6	8·1	2·2	14·1	3·9
22	4 05·5	4 06·2	3 54·3	2·2	0·6	8·2	2·3	14·2	3·9
23	4 05·8	4 06·4	3 54·6	2·3	0·6	8·3	2·3	14·3	3·9
24	4 06·0	4 06·7	3 54·8	2·4	0·7	8·4	2·3	14·4	4·0
25	4 06·3	4 06·9	3 55·0	2·5	0·7	8·5	2·3	14·5	4·0
26	4 06·5	4 07·2	3 55·3	2·6	0·7	8·6	2·4	14·6	4·0
27	4 06·8	4 07·4	3 55·5	2·7	0·7	8·7	2·4	14·7	4·0
28	4 07·0	4 07·7	3 55·7	2·8	0·8	8·8	2·4	14·8	4·1
29	4 07·3	4 07·9	3 56·0	2·9	0·8	8·9	2·4	14·9	4·1
30	4 07·5	4 08·2	3 56·2	3·0	0·8	9·0	2·5	15·0	4·1
31	4 07·8	4 08·4	3 56·5	3·1	0·9	9·1	2·5	15·1	4·2
32	4 08·0	4 08·7	3 56·7	3·2	0·9	9·2	2·5	15·2	4·2
33	4 08·3	4 08·9	3 56·9	3·3	0·9	9·3	2·6	15·3	4·2
34	4 08·5	4 09·2	3 57·2	3·4	0·9	9·4	2·6	15·4	4·2
35	4 08·8	4 09·4	3 57·4	3·5	1·0	9·5	2·6	15·5	4·3
36	4 09·0	4 09·7	3 57·7	3·6	1·0	9·6	2·6	15·6	4·3
37	4 09·3	4 09·9	3 57·9	3·7	1·0	9·7	2·7	15·7	4·3
38	4 09·5	4 10·2	3 58·1	3·8	1·0	9·8	2·7	15·8	4·3
39	4 09·8	4 10·4	3 58·4	3·9	1·1	9·9	2·7	15·9	4·4
40	4 10·0	4 10·7	3 58·6	4·0	1·1	10·0	2·8	16·0	4·4
41	4 10·3	4 10·9	3 58·8	4·1	1·1	10·1	2·8	16·1	4·4
42	4 10·5	4 11·2	3 59·1	4·2	1·2	10·2	2·8	16·2	4·5
43	4 10·8	4 11·4	3 59·3	4·3	1·2	10·3	2·8	16·3	4·5
44	4 11·0	4 11·7	3 59·6	4·4	1·2	10·4	2·9	16·4	4·5
45	4 11·3	4 11·9	3 59·8	4·5	1·2	10·5	2·9	16·5	4·5
46	4 11·5	4 12·2	4 00·0	4·6	1·3	10·6	2·9	16·6	4·6
47	4 11·8	4 12·4	4 00·3	4·7	1·3	10·7	2·9	16·7	4·6
48	4 12·0	4 12·7	4 00·5	4·8	1·3	10·8	3·0	16·8	4·6
49	4 12·3	4 12·9	4 00·8	4·9	1·3	10·9	3·0	16·9	4·6
50	4 12·5	4 13·2	4 01·0	5·0	1·4	11·0	3·0	17·0	4·7
51	4 12·8	4 13·4	4 01·2	5·1	1·4	11·1	3·1	17·1	4·7
52	4 13·0	4 13·7	4 01·5	5·2	1·4	11·2	3·1	17·2	4·7
53	4 13·3	4 13·9	4 01·7	5·3	1·5	11·3	3·1	17·3	4·8
54	4 13·5	4 14·2	4 02·0	5·4	1·5	11·4	3·1	17·4	4·8
55	4 13·8	4 14·4	4 02·2	5·5	1·5	11·5	3·2	17·5	4·8
56	4 14·0	4 14·7	4 02·4	5·6	1·5	11·6	3·2	17·6	4·8
57	4 14·3	4 14·9	4 02·7	5·7	1·6	11·7	3·2	17·7	4·9
58	4 14·5	4 15·2	4 02·9	5·8	1·6	11·8	3·2	17·8	4·9
59	4 14·8	4 15·4	4 03·1	5·9	1·6	11·9	3·3	17·9	4·9
60	4 15·0	4 15·7	4 03·4	6·0	1·7	12·0	3·3	18·0	5·0

17ᵐ s	SUN PLANETS ° ′	ARIES ° ′	MOON ° ′	v or d ′	Corrⁿ ′	v or d ′	Corrⁿ ′	v or d ′	Corrⁿ ′
00	4 15·0	4 15·7	4 03·4	0·0	0·0	6·0	1·8	12·0	3·5
01	4 15·3	4 15·9	4 03·6	0·1	0·0	6·1	1·8	12·1	3·5
02	4 15·5	4 16·2	4 03·9	0·2	0·1	6·2	1·8	12·2	3·6
03	4 15·8	4 16·5	4 04·1	0·3	0·1	6·3	1·8	12·3	3·6
04	4 16·0	4 16·7	4 04·3	0·4	0·1	6·4	1·9	12·4	3·6
05	4 16·3	4 17·0	4 04·6	0·5	0·1	6·5	1·9	12·5	3·6
06	4 16·5	4 17·2	4 04·8	0·6	0·2	6·6	1·9	12·6	3·7
07	4 16·8	4 17·5	4 05·1	0·7	0·2	6·7	2·0	12·7	3·7
08	4 17·0	4 17·7	4 05·3	0·8	0·2	6·8	2·0	12·8	3·7
09	4 17·3	4 18·0	4 05·5	0·9	0·3	6·9	2·0	12·9	3·8
10	4 17·5	4 18·2	4 05·8	1·0	0·3	7·0	2·0	13·0	3·8
11	4 17·8	4 18·5	4 06·0	1·1	0·3	7·1	2·1	13·1	3·8
12	4 18·0	4 18·7	4 06·2	1·2	0·4	7·2	2·1	13·2	3·9
13	4 18·3	4 19·0	4 06·5	1·3	0·4	7·3	2·1	13·3	3·9
14	4 18·5	4 19·2	4 06·7	1·4	0·4	7·4	2·2	13·4	3·9
15	4 18·8	4 19·5	4 07·0	1·5	0·4	7·5	2·2	13·5	3·9
16	4 19·0	4 19·7	4 07·2	1·6	0·5	7·6	2·2	13·6	4·0
17	4 19·3	4 20·0	4 07·4	1·7	0·5	7·7	2·2	13·7	4·0
18	4 19·5	4 20·2	4 07·7	1·8	0·5	7·8	2·3	13·8	4·0
19	4 19·8	4 20·5	4 07·9	1·9	0·6	7·9	2·3	13·9	4·1
20	4 20·0	4 20·7	4 08·2	2·0	0·6	8·0	2·3	14·0	4·1
21	4 20·3	4 21·0	4 08·4	2·1	0·6	8·1	2·4	14·1	4·1
22	4 20·5	4 21·2	4 08·6	2·2	0·6	8·2	2·4	14·2	4·1
23	4 20·8	4 21·5	4 08·9	2·3	0·7	8·3	2·4	14·3	4·2
24	4 21·0	4 21·7	4 09·1	2·4	0·7	8·4	2·5	14·4	4·2
25	4 21·3	4 22·0	4 09·3	2·5	0·7	8·5	2·5	14·5	4·2
26	4 21·5	4 22·2	4 09·6	2·6	0·8	8·6	2·5	14·6	4·3
27	4 21·8	4 22·5	4 09·8	2·7	0·8	8·7	2·5	14·7	4·3
28	4 22·0	4 22·7	4 10·1	2·8	0·8	8·8	2·6	14·8	4·3
29	4 22·3	4 23·0	4 10·3	2·9	0·8	8·9	2·6	14·9	4·3
30	4 22·5	4 23·2	4 10·5	3·0	0·9	9·0	2·6	15·0	4·4
31	4 22·8	4 23·5	4 10·8	3·1	0·9	9·1	2·7	15·1	4·4
32	4 23·0	4 23·7	4 11·0	3·2	0·9	9·2	2·7	15·2	4·4
33	4 23·3	4 24·0	4 11·3	3·3	1·0	9·3	2·7	15·3	4·5
34	4 23·5	4 24·2	4 11·5	3·4	1·0	9·4	2·7	15·4	4·5
35	4 23·8	4 24·5	4 11·7	3·5	1·0	9·5	2·8	15·5	4·5
36	4 24·0	4 24·7	4 12·0	3·6	1·1	9·6	2·8	15·6	4·6
37	4 24·3	4 25·0	4 12·2	3·7	1·1	9·7	2·8	15·7	4·6
38	4 24·5	4 25·2	4 12·5	3·8	1·1	9·8	2·9	15·8	4·6
39	4 24·8	4 25·5	4 12·7	3·9	1·1	9·9	2·9	15·9	4·6
40	4 25·0	4 25·7	4 12·9	4·0	1·2	10·0	2·9	16·0	4·7
41	4 25·3	4 26·0	4 13·2	4·1	1·2	10·1	2·9	16·1	4·7
42	4 25·5	4 26·2	4 13·4	4·2	1·2	10·2	3·0	16·2	4·7
43	4 25·8	4 26·5	4 13·6	4·3	1·3	10·3	3·0	16·3	4·8
44	4 26·0	4 26·7	4 13·9	4·4	1·3	10·4	3·0	16·4	4·8
45	4 26·3	4 27·0	4 14·1	4·5	1·3	10·5	3·1	16·5	4·8
46	4 26·5	4 27·2	4 14·4	4·6	1·3	10·6	3·1	16·6	4·8
47	4 26·8	4 27·5	4 14·6	4·7	1·4	10·7	3·1	16·7	4·9
48	4 27·0	4 27·7	4 14·8	4·8	1·4	10·8	3·1	16·8	4·9
49	4 27·3	4 28·0	4 15·1	4·9	1·4	10·9	3·2	16·9	4·9
50	4 27·5	4 28·2	4 15·3	5·0	1·5	11·0	3·2	17·0	5·0
51	4 27·8	4 28·5	4 15·6	5·1	1·5	11·1	3·2	17·1	5·0
52	4 28·0	4 28·7	4 15·8	5·2	1·5	11·2	3·3	17·2	5·0
53	4 28·3	4 29·0	4 16·0	5·3	1·5	11·3	3·3	17·3	5·0
54	4 28·5	4 29·2	4 16·3	5·4	1·6	11·4	3·3	17·4	5·1
55	4 28·8	4 29·5	4 16·5	5·5	1·6	11·5	3·4	17·5	5·1
56	4 29·0	4 29·7	4 16·7	5·6	1·6	11·6	3·4	17·6	5·1
57	4 29·3	4 30·0	4 17·0	5·7	1·7	11·7	3·4	17·7	5·2
58	4 29·5	4 30·2	4 17·2	5·8	1·7	11·8	3·4	17·8	5·2
59	4 29·8	4 30·5	4 17·5	5·9	1·7	11·9	3·5	17·9	5·2
60	4 30·0	4 30·7	4 17·7	6·0	1·8	12·0	3·5	18·0	5·3

18ᵐ	SUN PLANETS	ARIES	MOON	v or Corrⁿ d		v or Corrⁿ d		v or Corrⁿ d	
s	° ′	° ′	° ′	′	′	′	′	′	′
00	4 30·0	4 30·7	4 17·7	0·0	0·0	6·0	1·9	12·0	3·7
01	4 30·3	4 31·0	4 17·9	0·1	0·0	6·1	1·9	12·1	3·7
02	4 30·5	4 31·2	4 18·2	0·2	0·1	6·2	1·9	12·2	3·8
03	4 30·8	4 31·5	4 18·4	0·3	0·1	6·3	1·9	12·3	3·8
04	4 31·0	4 31·7	4 18·7	0·4	0·1	6·4	2·0	12·4	3·8
05	4 31·3	4 32·0	4 18·9	0·5	0·2	6·5	2·0	12·5	3·9
06	4 31·5	4 32·2	4 19·1	0·6	0·2	6·6	2·0	12·6	3·9
07	4 31·8	4 32·5	4 19·4	0·7	0·2	6·7	2·1	12·7	3·9
08	4 32·0	4 32·7	4 19·6	0·8	0·2	6·8	2·1	12·8	3·9
09	4 32·3	4 33·0	4 19·8	0·9	0·3	6·9	2·1	12·9	4·0
10	4 32·5	4 33·2	4 20·1	1·0	0·3	7·0	2·2	13·0	4·0
11	4 32·8	4 33·5	4 20·3	1·1	0·3	7·1	2·2	13·1	4·0
12	4 33·0	4 33·7	4 20·6	1·2	0·4	7·2	2·2	13·2	4·1
13	4 33·3	4 34·0	4 20·8	1·3	0·4	7·3	2·3	13·3	4·1
14	4 33·5	4 34·2	4 21·0	1·4	0·4	7·4	2·3	13·4	4·1
15	4 33·8	4 34·5	4 21·3	1·5	0·5	7·5	2·3	13·5	4·2
16	4 34·0	4 34·8	4 21·5	1·6	0·5	7·6	2·3	13·6	4·2
17	4 34·3	4 35·0	4 21·8	1·7	0·5	7·7	2·4	13·7	4·2
18	4 34·5	4 35·3	4 22·0	1·8	0·6	7·8	2·4	13·8	4·3
19	4 34·8	4 35·5	4 22·2	1·9	0·6	7·9	2·4	13·9	4·3
20	4 35·0	4 35·8	4 22·5	2·0	0·6	8·0	2·5	14·0	4·3
21	4 35·3	4 36·0	4 22·7	2·1	0·6	8·1	2·5	14·1	4·3
22	4 35·5	4 36·3	4 22·9	2·2	0·7	8·2	2·5	14·2	4·4
23	4 35·8	4 36·5	4 23·2	2·3	0·7	8·3	2·6	14·3	4·4
24	4 36·0	4 36·8	4 23·4	2·4	0·7	8·4	2·6	14·4	4·4
25	4 36·3	4 37·0	4 23·7	2·5	0·8	8·5	2·6	14·5	4·5
26	4 36·5	4 37·3	4 23·9	2·6	0·8	8·6	2·7	14·6	4·5
27	4 36·8	4 37·5	4 24·1	2·7	0·8	8·7	2·7	14·7	4·5
28	4 37·0	4 37·8	4 24·4	2·8	0·9	8·8	2·7	14·8	4·6
29	4 37·3	4 38·0	4 24·6	2·9	0·9	8·9	2·7	14·9	4·6
30	4 37·5	4 38·3	4 24·9	3·0	0·9	9·0	2·8	15·0	4·6
31	4 37·8	4 38·5	4 25·1	3·1	1·0	9·1	2·8	15·1	4·7
32	4 38·0	4 38·8	4 25·3	3·2	1·0	9·2	2·8	15·2	4·7
33	4 38·3	4 39·0	4 25·6	3·3	1·0	9·3	2·9	15·3	4·7
34	4 38·5	4 39·3	4 25·8	3·4	1·0	9·4	2·9	15·4	4·7
35	4 38·8	4 39·5	4 26·1	3·5	1·1	9·5	2·9	15·5	4·8
36	4 39·0	4 39·8	4 26·3	3·6	1·1	9·6	3·0	15·6	4·8
37	4 39·3	4 40·0	4 26·5	3·7	1·1	9·7	3·0	15·7	4·8
38	4 39·5	4 40·3	4 26·8	3·8	1·2	9·8	3·0	15·8	4·9
39	4 39·8	4 40·5	4 27·0	3·9	1·2	9·9	3·1	15·9	4·9
40	4 40·0	4 40·8	4 27·2	4·0	1·2	10·0	3·1	16·0	4·9
41	4 40·3	4 41·0	4 27·5	4·1	1·3	10·1	3·1	16·1	5·0
42	4 40·5	4 41·3	4 27·7	4·2	1·3	10·2	3·1	16·2	5·0
43	4 40·8	4 41·5	4 28·0	4·3	1·3	10·3	3·2	16·3	5·0
44	4 41·0	4 41·8	4 28·2	4·4	1·4	10·4	3·2	16·4	5·1
45	4 41·3	4 42·0	4 28·4	4·5	1·4	10·5	3·2	16·5	5·1
46	4 41·5	4 42·3	4 28·7	4·6	1·4	10·6	3·3	16·6	5·1
47	4 41·8	4 42·5	4 28·9	4·7	1·4	10·7	3·3	16·7	5·1
48	4 42·0	4 42·8	4 29·2	4·8	1·5	10·8	3·3	16·8	5·2
49	4 42·3	4 43·0	4 29·4	4·9	1·5	10·9	3·4	16·9	5·2
50	4 42·5	4 43·3	4 29·6	5·0	1·5	11·0	3·4	17·0	5·2
51	4 42·8	4 43·5	4 29·9	5·1	1·6	11·1	3·4	17·1	5·3
52	4 43·0	4 43·8	4 30·1	5·2	1·6	11·2	3·5	17·2	5·3
53	4 43·3	4 44·0	4 30·3	5·3	1·6	11·3	3·5	17·3	5·3
54	4 43·5	4 44·3	4 30·6	5·4	1·7	11·4	3·5	17·4	5·4
55	4 43·8	4 44·5	4 30·8	5·5	1·7	11·5	3·5	17·5	5·4
56	4 44·0	4 44·8	4 31·1	5·6	1·7	11·6	3·6	17·6	5·4
57	4 44·3	4 45·0	4 31·3	5·7	1·8	11·7	3·6	17·7	5·5
58	4 44·5	4 45·3	4 31·5	5·8	1·8	11·8	3·6	17·8	5·5
59	4 44·8	4 45·5	4 31·8	5·9	1·8	11·9	3·7	17·9	5·5
60	4 45·0	4 45·8	4 32·0	6·0	1·9	12·0	3·7	18·0	5·6

19ᵐ	SUN PLANETS	ARIES	MOON	v or Corrⁿ d		v or Corrⁿ d		v or Corrⁿ d	
s	° ′	° ′	° ′	′	′	′	′	′	′
00	4 45·0	4 45·8	4 32·0	0·0	0·0	6·0	2·0	12·0	3·9
01	4 45·3	4 46·0	4 32·3	0·1	0·0	6·1	2·0	12·1	3·9
02	4 45·5	4 46·3	4 32·5	0·2	0·1	6·2	2·0	12·2	4·0
03	4 45·8	4 46·5	4 32·7	0·3	0·1	6·3	2·0	12·3	4·0
04	4 46·0	4 46·8	4 33·0	0·4	0·1	6·4	2·1	12·4	4·0
05	4 46·3	4 47·0	4 33·2	0·5	0·2	6·5	2·1	12·5	4·1
06	4 46·5	4 47·3	4 33·4	0·6	0·2	6·6	2·1	12·6	4·1
07	4 46·8	4 47·5	4 33·7	0·7	0·2	6·7	2·2	12·7	4·1
08	4 47·0	4 47·8	4 33·9	0·8	0·3	6·8	2·2	12·8	4·2
09	4 47·3	4 48·0	4 34·2	0·9	0·3	6·9	2·2	12·9	4·2
10	4 47·5	4 48·3	4 34·4	1·0	0·3	7·0	2·3	13·0	4·2
11	4 47·8	4 48·5	4 34·6	1·1	0·4	7·1	2·3	13·1	4·3
12	4 48·0	4 48·8	4 34·9	1·2	0·4	7·2	2·3	13·2	4·3
13	4 48·3	4 49·0	4 35·1	1·3	0·4	7·3	2·4	13·3	4·3
14	4 48·5	4 49·3	4 35·4	1·4	0·5	7·4	2·4	13·4	4·4
15	4 48·8	4 49·5	4 35·6	1·5	0·5	7·5	2·4	13·5	4·4
16	4 49·0	4 49·8	4 35·8	1·6	0·5	7·6	2·5	13·6	4·4
17	4 49·3	4 50·0	4 36·1	1·7	0·6	7·7	2·5	13·7	4·5
18	4 49·5	4 50·3	4 36·3	1·8	0·6	7·8	2·5	13·8	4·5
19	4 49·8	4 50·5	4 36·6	1·9	0·6	7·9	2·6	13·9	4·5
20	4 50·0	4 50·8	4 36·8	2·0	0·7	8·0	2·6	14·0	4·6
21	4 50·3	4 51·0	4 37·0	2·1	0·7	8·1	2·6	14·1	4·6
22	4 50·5	4 51·3	4 37·3	2·2	0·7	8·2	2·7	14·2	4·6
23	4 50·8	4 51·5	4 37·5	2·3	0·7	8·3	2·7	14·3	4·6
24	4 51·0	4 51·8	4 37·7	2·4	0·8	8·4	2·7	14·4	4·7
25	4 51·3	4 52·0	4 38·0	2·5	0·8	8·5	2·8	14·5	4·7
26	4 51·5	4 52·3	4 38·2	2·6	0·8	8·6	2·8	14·6	4·7
27	4 51·8	4 52·5	4 38·5	2·7	0·9	8·7	2·8	14·7	4·8
28	4 52·0	4 52·8	4 38·7	2·8	0·9	8·8	2·9	14·8	4·8
29	4 52·3	4 53·1	4 38·9	2·9	0·9	8·9	2·9	14·9	4·8
30	4 52·5	4 53·3	4 39·2	3·0	1·0	9·0	2·9	15·0	4·9
31	4 52·8	4 53·6	4 39·4	3·1	1·0	9·1	3·0	15·1	4·9
32	4 53·0	4 53·8	4 39·7	3·2	1·0	9·2	3·0	15·2	4·9
33	4 53·3	4 54·1	4 39·9	3·3	1·1	9·3	3·0	15·3	5·0
34	4 53·5	4 54·3	4 40·1	3·4	1·1	9·4	3·1	15·4	5·0
35	4 53·8	4 54·6	4 40·4	3·5	1·1	9·5	3·1	15·5	5·0
36	4 54·0	4 54·8	4 40·6	3·6	1·2	9·6	3·1	15·6	5·1
37	4 54·3	4 55·1	4 40·8	3·7	1·2	9·7	3·2	15·7	5·1
38	4 54·5	4 55·3	4 41·1	3·8	1·2	9·8	3·2	15·8	5·1
39	4 54·8	4 55·6	4 41·3	3·9	1·3	9·9	3·2	15·9	5·2
40	4 55·0	4 55·8	4 41·6	4·0	1·3	10·0	3·3	16·0	5·2
41	4 55·3	4 56·1	4 41·8	4·1	1·3	10·1	3·3	16·1	5·2
42	4 55·5	4 56·3	4 42·0	4·2	1·4	10·2	3·3	16·2	5·3
43	4 55·8	4 56·6	4 42·3	4·3	1·4	10·3	3·3	16·3	5·3
44	4 56·0	4 56·8	4 42·5	4·4	1·4	10·4	3·4	16·4	5·3
45	4 56·3	4 57·1	4 42·8	4·5	1·5	10·5	3·4	16·5	5·4
46	4 56·5	4 57·3	4 43·0	4·6	1·5	10·6	3·4	16·6	5·4
47	4 56·8	4 57·6	4 43·2	4·7	1·5	10·7	3·5	16·7	5·4
48	4 57·0	4 57·8	4 43·5	4·8	1·6	10·8	3·5	16·8	5·5
49	4 57·3	4 58·1	4 43·7	4·9	1·6	10·9	3·5	16·9	5·5
50	4 57·5	4 58·3	4 43·9	5·0	1·6	11·0	3·6	17·0	5·5
51	4 57·8	4 58·6	4 44·2	5·1	1·7	11·1	3·6	17·1	5·6
52	4 58·0	4 58·8	4 44·4	5·2	1·7	11·2	3·6	17·2	5·6
53	4 58·3	4 59·1	4 44·7	5·3	1·7	11·3	3·7	17·3	5·6
54	4 58·5	4 59·3	4 44·9	5·4	1·8	11·4	3·7	17·4	5·7
55	4 58·8	4 59·6	4 45·1	5·5	1·8	11·5	3·7	17·5	5·7
56	4 59·0	4 59·8	4 45·4	5·6	1·8	11·6	3·8	17·6	5·7
57	4 59·3	5 00·1	4 45·6	5·7	1·9	11·7	3·8	17·7	5·8
58	4 59·5	5 00·3	4 45·9	5·8	1·9	11·8	3·8	17·8	5·8
59	4 59·8	5 00·6	4 46·1	5·9	1·9	11·9	3·9	17·9	5·8
60	5 00·0	5 00·8	4 46·3	6·0	2·0	12·0	3·9	18·0	5·9

20ᵐ

20ᵐ s	SUN PLANETS	ARIES	MOON	v or Corrⁿ d	v or Corrⁿ d	v or Corrⁿ d
00	5 00·0	5 00·8	4 46·3	0·0 0·0	6·0 2·1	12·0 4·1
01	5 00·3	5 01·1	4 46·6	0·1 0·0	6·1 2·1	12·1 4·1
02	5 00·5	5 01·3	4 46·8	0·2 0·1	6·2 2·1	12·2 4·2
03	5 00·8	5 01·6	4 47·0	0·3 0·1	6·3 2·2	12·3 4·2
04	5 01·0	5 01·8	4 47·3	0·4 0·1	6·4 2·2	12·4 4·2
05	5 01·3	5 02·1	4 47·5	0·5 0·2	6·5 2·2	12·5 4·3
06	5 01·5	5 02·3	4 47·8	0·6 0·2	6·6 2·3	12·6 4·3
07	5 01·8	5 02·6	4 48·0	0·7 0·2	6·7 2·3	12·7 4·3
08	5 02·0	5 02·8	4 48·2	0·8 0·3	6·8 2·3	12·8 4·4
09	5 02·3	5 03·1	4 48·5	0·9 0·3	6·9 2·4	12·9 4·4
10	5 02·5	5 03·3	4 48·7	1·0 0·3	7·0 2·4	13·0 4·4
11	5 02·8	5 03·6	4 49·0	1·1 0·4	7·1 2·4	13·1 4·5
12	5 03·0	5 03·8	4 49·2	1·2 0·4	7·2 2·5	13·2 4·5
13	5 03·3	5 04·1	4 49·4	1·3 0·4	7·3 2·5	13·3 4·5
14	5 03·5	5 04·3	4 49·7	1·4 0·5	7·4 2·5	13·4 4·6
15	5 03·8	5 04·6	4 49·9	1·5 0·5	7·5 2·6	13·5 4·6
16	5 04·0	5 04·8	4 50·2	1·6 0·5	7·6 2·6	13·6 4·6
17	5 04·3	5 05·1	4 50·4	1·7 0·6	7·7 2·6	13·7 4·7
18	5 04·5	5 05·3	4 50·6	1·8 0·6	7·8 2·7	13·8 4·7
19	5 04·8	5 05·6	4 50·9	1·9 0·6	7·9 2·7	13·9 4·7
20	5 05·0	5 05·8	4 51·1	2·0 0·7	8·0 2·7	14·0 4·8
21	5 05·3	5 06·1	4 51·3	2·1 0·7	8·1 2·8	14·1 4·8
22	5 05·5	5 06·3	4 51·6	2·2 0·8	8·2 2·8	14·2 4·9
23	5 05·8	5 06·6	4 51·8	2·3 0·8	8·3 2·8	14·3 4·9
24	5 06·0	5 06·8	4 52·1	2·4 0·8	8·4 2·9	14·4 4·9
25	5 06·3	5 07·1	4 52·3	2·5 0·9	8·5 2·9	14·5 5·0
26	5 06·5	5 07·3	4 52·5	2·6 0·9	8·6 2·9	14·6 5·0
27	5 06·8	5 07·6	4 52·8	2·7 0·9	8·7 3·0	14·7 5·0
28	5 07·0	5 07·8	4 53·0	2·8 1·0	8·8 3·0	14·8 5·1
29	5 07·3	5 08·1	4 53·3	2·9 1·0	8·9 3·0	14·9 5·1
30	5 07·5	5 08·3	4 53·5	3·0 1·0	9·0 3·1	15·0 5·1
31	5 07·8	5 08·6	4 53·7	3·1 1·1	9·1 3·1	15·1 5·2
32	5 08·0	5 08·8	4 54·0	3·2 1·1	9·2 3·1	15·2 5·2
33	5 08·3	5 09·1	4 54·2	3·3 1·1	9·3 3·2	15·3 5·2
34	5 08·5	5 09·3	4 54·4	3·4 1·2	9·4 3·2	15·4 5·3
35	5 08·8	5 09·6	4 54·7	3·5 1·2	9·5 3·2	15·5 5·3
36	5 09·0	5 09·8	4 54·9	3·6 1·2	9·6 3·3	15·6 5·3
37	5 09·3	5 10·1	4 55·2	3·7 1·3	9·7 3·3	15·7 5·4
38	5 09·5	5 10·3	4 55·4	3·8 1·3	9·8 3·3	15·8 5·4
39	5 09·8	5 10·6	4 55·6	3·9 1·3	9·9 3·4	15·9 5·4
40	5 10·0	5 10·8	4 55·9	4·0 1·4	10·0 3·4	16·0 5·5
41	5 10·3	5 11·1	4 56·1	4·1 1·4	10·1 3·5	16·1 5·5
42	5 10·5	5 11·4	4 56·4	4·2 1·4	10·2 3·5	16·2 5·5
43	5 10·8	5 11·6	4 56·6	4·3 1·5	10·3 3·5	16·3 5·6
44	5 11·0	5 11·9	4 56·8	4·4 1·5	10·4 3·6	16·4 5·6
45	5 11·3	5 12·1	4 57·1	4·5 1·5	10·5 3·6	16·5 5·6
46	5 11·5	5 12·4	4 57·3	4·6 1·6	10·6 3·6	16·6 5·7
47	5 11·8	5 12·6	4 57·5	4·7 1·6	10·7 3·7	16·7 5·7
48	5 12·0	5 12·9	4 57·8	4·8 1·6	10·8 3·7	16·8 5·7
49	5 12·3	5 13·1	4 58·0	4·9 1·7	10·9 3·7	16·9 5·8
50	5 12·5	5 13·4	4 58·3	5·0 1·7	11·0 3·8	17·0 5·8
51	5 12·8	5 13·6	4 58·5	5·1 1·7	11·1 3·8	17·1 5·8
52	5 13·0	5 13·9	4 58·7	5·2 1·8	11·2 3·8	17·2 5·9
53	5 13·3	5 14·1	4 59·0	5·3 1·8	11·3 3·9	17·3 5·9
54	5 13·5	5 14·4	4 59·2	5·4 1·8	11·4 3·9	17·4 5·9
55	5 13·8	5 14·6	4 59·5	5·5 1·9	11·5 3·9	17·5 6·0
56	5 14·0	5 14·9	4 59·7	5·6 1·9	11·6 4·0	17·6 6·0
57	5 14·3	5 15·1	4 59·9	5·7 1·9	11·7 4·0	17·7 6·0
58	5 14·5	5 15·4	5 00·2	5·8 2·0	11·8 4·0	17·8 6·1
59	5 14·8	5 15·6	5 00·4	5·9 2·0	11·9 4·1	17·9 6·1
60	5 15·0	5 15·9	5 00·7	6·0 2·1	12·0 4·1	18·0 6·2

21ᵐ

21ᵐ s	SUN PLANETS	ARIES	MOON	v or Corrⁿ d	v or Corrⁿ d	v or Corrⁿ d
00	5 15·0	5 15·9	5 00·7	0·0 0·0	6·0 2·2	12·0 4·3
01	5 15·3	5 16·1	5 00·9	0·1 0·0	6·1 2·2	12·1 4·3
02	5 15·5	5 16·4	5 01·1	0·2 0·1	6·2 2·2	12·2 4·4
03	5 15·8	5 16·6	5 01·4	0·3 0·1	6·3 2·3	12·3 4·4
04	5 16·0	5 16·9	5 01·6	0·4 0·1	6·4 2·3	12·4 4·4
05	5 16·3	5 17·1	5 01·8	0·5 0·2	6·5 2·3	12·5 4·5
06	5 16·5	5 17·4	5 02·1	0·6 0·2	6·6 2·4	12·6 4·5
07	5 16·8	5 17·6	5 02·3	0·7 0·3	6·7 2·4	12·7 4·6
08	5 17·0	5 17·9	5 02·6	0·8 0·3	6·8 2·4	12·8 4·6
09	5 17·3	5 18·1	5 02·8	0·9 0·3	6·9 2·5	12·9 4·6
10	5 17·5	5 18·4	5 03·0	1·0 0·4	7·0 2·5	13·0 4·7
11	5 17·8	5 18·6	5 03·3	1·1 0·4	7·1 2·5	13·1 4·7
12	5 18·0	5 18·9	5 03·5	1·2 0·4	7·2 2·6	13·2 4·7
13	5 18·3	5 19·1	5 03·8	1·3 0·5	7·3 2·6	13·3 4·8
14	5 18·5	5 19·4	5 04·0	1·4 0·5	7·4 2·7	13·4 4·8
15	5 18·8	5 19·6	5 04·2	1·5 0·5	7·5 2·7	13·5 4·8
16	5 19·0	5 19·9	5 04·5	1·6 0·6	7·6 2·7	13·6 4·9
17	5 19·3	5 20·1	5 04·7	1·7 0·6	7·7 2·8	13·7 4·9
18	5 19·5	5 20·4	5 04·9	1·8 0·6	7·8 2·8	13·8 4·9
19	5 19·8	5 20·6	5 05·2	1·9 0·7	7·9 2·8	13·9 5·0
20	5 20·0	5 20·9	5 05·4	2·0 0·7	8·0 2·9	14·0 5·0
21	5 20·3	5 21·1	5 05·7	2·1 0·8	8·1 2·9	14·1 5·1
22	5 20·5	5 21·4	5 05·9	2·2 0·8	8·2 2·9	14·2 5·1
23	5 20·8	5 21·6	5 06·1	2·3 0·8	8·3 3·0	14·3 5·1
24	5 21·0	5 21·9	5 06·4	2·4 0·9	8·4 3·0	14·4 5·2
25	5 21·3	5 22·1	5 06·6	2·5 0·9	8·5 3·0	14·5 5·2
26	5 21·5	5 22·4	5 06·9	2·6 0·9	8·6 3·1	14·6 5·2
27	5 21·8	5 22·6	5 07·1	2·7 1·0	8·7 3·1	14·7 5·3
28	5 22·0	5 22·9	5 07·3	2·8 1·0	8·8 3·2	14·8 5·3
29	5 22·3	5 23·1	5 07·6	2·9 1·0	8·9 3·2	14·9 5·3
30	5 22·5	5 23·4	5 07·8	3·0 1·1	9·0 3·2	15·0 5·4
31	5 22·8	5 23·6	5 08·0	3·1 1·1	9·1 3·3	15·1 5·4
32	5 23·0	5 23·9	5 08·3	3·2 1·1	9·2 3·3	15·2 5·4
33	5 23·3	5 24·1	5 08·5	3·3 1·2	9·3 3·3	15·3 5·5
34	5 23·5	5 24·4	5 08·8	3·4 1·2	9·4 3·4	15·4 5·5
35	5 23·8	5 24·6	5 09·0	3·5 1·3	9·5 3·4	15·5 5·6
36	5 24·0	5 24·9	5 09·2	3·6 1·3	9·6 3·4	15·6 5·6
37	5 24·3	5 25·1	5 09·5	3·7 1·3	9·7 3·5	15·7 5·6
38	5 24·5	5 25·4	5 09·7	3·8 1·4	9·8 3·5	15·8 5·7
39	5 24·8	5 25·6	5 10·0	3·9 1·4	9·9 3·5	15·9 5·7
40	5 25·0	5 25·9	5 10·2	4·0 1·4	10·0 3·6	16·0 5·7
41	5 25·3	5 26·1	5 10·4	4·1 1·5	10·1 3·6	16·1 5·8
42	5 25·5	5 26·4	5 10·7	4·2 1·5	10·2 3·7	16·2 5·8
43	5 25·8	5 26·6	5 10·9	4·3 1·5	10·3 3·7	16·3 5·8
44	5 26·0	5 26·9	5 11·1	4·4 1·6	10·4 3·7	16·4 5·9
45	5 26·3	5 27·1	5 11·4	4·5 1·6	10·5 3·8	16·5 5·9
46	5 26·5	5 27·4	5 11·6	4·6 1·6	10·6 3·8	16·6 5·9
47	5 26·8	5 27·6	5 11·9	4·7 1·7	10·7 3·8	16·7 6·0
48	5 27·0	5 27·9	5 12·1	4·8 1·7	10·8 3·9	16·8 6·0
49	5 27·3	5 28·1	5 12·3	4·9 1·8	10·9 3·9	16·9 6·1
50	5 27·5	5 28·4	5 12·6	5·0 1·8	11·0 3·9	17·0 6·1
51	5 27·8	5 28·6	5 12·8	5·1 1·8	11·1 4·0	17·1 6·1
52	5 28·0	5 28·9	5 13·1	5·2 1·9	11·2 4·0	17·2 6·2
53	5 28·3	5 29·1	5 13·3	5·3 1·9	11·3 4·0	17·3 6·2
54	5 28·5	5 29·4	5 13·5	5·4 1·9	11·4 4·1	17·4 6·2
55	5 28·8	5 29·7	5 13·8	5·5 2·0	11·5 4·1	17·5 6·3
56	5 29·0	5 29·9	5 14·0	5·6 2·0	11·6 4·2	17·6 6·3
57	5 29·3	5 30·2	5 14·3	5·7 2·0	11·7 4·2	17·7 6·3
58	5 29·5	5 30·4	5 14·5	5·8 2·1	11·8 4·2	17·8 6·4
59	5 29·8	5 30·7	5 14·7	5·9 2·1	11·9 4·3	17·9 6·4
60	5 30·0	5 30·9	5 15·0	6·0 2·2	12·0 4·3	18·0 6·5

22ᵐ

22ᵐ s	SUN PLANETS ° ′	ARIES ° ′	MOON ° ′	v or d ′	Corrⁿ ′	v or d ′	Corrⁿ ′	v or d ′	Corrⁿ ′
00	5 30.0	5 30.9	5 15.0	0.0	0.0	6.0	2.3	12.0	4.5
01	5 30.3	5 31.2	5 15.2	0.1	0.0	6.1	2.3	12.1	4.5
02	5 30.5	5 31.4	5 15.4	0.2	0.1	6.2	2.3	12.2	4.6
03	5 30.8	5 31.7	5 15.7	0.3	0.1	6.3	2.4	12.3	4.6
04	5 31.0	5 31.9	5 15.9	0.4	0.2	6.4	2.4	12.4	4.7
05	5 31.3	5 32.2	5 16.2	0.5	0.2	6.5	2.4	12.5	4.7
06	5 31.5	5 32.4	5 16.4	0.6	0.2	6.6	2.5	12.6	4.7
07	5 31.8	5 32.7	5 16.6	0.7	0.3	6.7	2.5	12.7	4.8
08	5 32.0	5 32.9	5 16.9	0.8	0.3	6.8	2.6	12.8	4.8
09	5 32.3	5 33.2	5 17.1	0.9	0.3	6.9	2.6	12.9	4.8
10	5 32.5	5 33.4	5 17.4	1.0	0.4	7.0	2.6	13.0	4.9
11	5 32.8	5 33.7	5 17.6	1.1	0.4	7.1	2.7	13.1	4.9
12	5 33.0	5 33.9	5 17.8	1.2	0.5	7.2	2.7	13.2	5.0
13	5 33.3	5 34.2	5 18.1	1.3	0.5	7.3	2.7	13.3	5.0
14	5 33.5	5 34.4	5 18.3	1.4	0.5	7.4	2.8	13.4	5.0
15	5 33.8	5 34.7	5 18.5	1.5	0.6	7.5	2.8	13.5	5.1
16	5 34.0	5 34.9	5 18.8	1.6	0.6	7.6	2.9	13.6	5.1
17	5 34.3	5 35.2	5 19.0	1.7	0.6	7.7	2.9	13.7	5.1
18	5 34.5	5 35.4	5 19.3	1.8	0.7	7.8	2.9	13.8	5.2
19	5 34.8	5 35.7	5 19.5	1.9	0.7	7.9	3.0	13.9	5.2
20	5 35.0	5 35.9	5 19.7	2.0	0.8	8.0	3.0	14.0	5.3
21	5 35.3	5 36.2	5 20.0	2.1	0.8	8.1	3.0	14.1	5.3
22	5 35.5	5 36.4	5 20.2	2.2	0.8	8.2	3.1	14.2	5.3
23	5 35.8	5 36.7	5 20.5	2.3	0.9	8.3	3.1	14.3	5.4
24	5 36.0	5 36.9	5 20.7	2.4	0.9	8.4	3.2	14.4	5.4
25	5 36.3	5 37.2	5 20.9	2.5	0.9	8.5	3.2	14.5	5.4
26	5 36.5	5 37.4	5 21.2	2.6	1.0	8.6	3.2	14.6	5.5
27	5 36.8	5 37.7	5 21.4	2.7	1.0	8.7	3.3	14.7	5.5
28	5 37.0	5 37.9	5 21.6	2.8	1.1	8.8	3.3	14.8	5.6
29	5 37.3	5 38.2	5 21.9	2.9	1.1	8.9	3.3	14.9	5.6
30	5 37.5	5 38.4	5 22.1	3.0	1.1	9.0	3.4	15.0	5.6
31	5 37.8	5 38.7	5 22.4	3.1	1.2	9.1	3.4	15.1	5.7
32	5 38.0	5 38.9	5 22.6	3.2	1.2	9.2	3.5	15.2	5.7
33	5 38.3	5 39.2	5 22.8	3.3	1.2	9.3	3.5	15.3	5.7
34	5 38.5	5 39.4	5 23.1	3.4	1.3	9.4	3.5	15.4	5.8
35	5 38.8	5 39.7	5 23.3	3.5	1.3	9.5	3.6	15.5	5.8
36	5 39.0	5 39.9	5 23.6	3.6	1.4	9.6	3.6	15.6	5.9
37	5 39.3	5 40.2	5 23.8	3.7	1.4	9.7	3.6	15.7	5.9
38	5 39.5	5 40.4	5 24.0	3.8	1.4	9.8	3.7	15.8	5.9
39	5 39.8	5 40.7	5 24.3	3.9	1.5	9.9	3.7	15.9	6.0
40	5 40.0	5 40.9	5 24.5	4.0	1.5	10.0	3.8	16.0	6.0
41	5 40.3	5 41.2	5 24.7	4.1	1.5	10.1	3.8	16.1	6.0
42	5 40.5	5 41.4	5 25.0	4.2	1.6	10.2	3.8	16.2	6.1
43	5 40.8	5 41.7	5 25.2	4.3	1.6	10.3	3.9	16.3	6.1
44	5 41.0	5 41.9	5 25.5	4.4	1.7	10.4	3.9	16.4	6.2
45	5 41.3	5 42.2	5 25.7	4.5	1.7	10.5	3.9	16.5	6.2
46	5 41.5	5 42.4	5 25.9	4.6	1.7	10.6	4.0	16.6	6.2
47	5 41.8	5 42.7	5 26.2	4.7	1.8	10.7	4.0	16.7	6.3
48	5 42.0	5 42.9	5 26.4	4.8	1.8	10.8	4.1	16.8	6.3
49	5 42.3	5 43.2	5 26.7	4.9	1.8	10.9	4.1	16.9	6.3
50	5 42.5	5 43.4	5 26.9	5.0	1.9	11.0	4.1	17.0	6.4
51	5 42.8	5 43.7	5 27.1	5.1	1.9	11.1	4.2	17.1	6.4
52	5 43.0	5 43.9	5 27.4	5.2	2.0	11.2	4.2	17.2	6.5
53	5 43.3	5 44.2	5 27.6	5.3	2.0	11.3	4.2	17.3	6.5
54	5 43.5	5 44.4	5 27.9	5.4	2.0	11.4	4.3	17.4	6.5
55	5 43.8	5 44.7	5 28.1	5.5	2.1	11.5	4.3	17.5	6.6
56	5 44.0	5 44.9	5 28.3	5.6	2.1	11.6	4.4	17.6	6.6
57	5 44.3	5 45.2	5 28.6	5.7	2.1	11.7	4.4	17.7	6.6
58	5 44.5	5 45.4	5 28.8	5.8	2.2	11.8	4.4	17.8	6.7
59	5 44.8	5 45.7	5 29.0	5.9	2.2	11.9	4.5	17.9	6.7
60	5 45.0	5 45.9	5 29.3	6.0	2.3	12.0	4.5	18.0	6.8

23ᵐ

23ᵐ s	SUN PLANETS ° ′	ARIES ° ′	MOON ° ′	v or d ′	Corrⁿ ′	v or d ′	Corrⁿ ′	v or d ′	Corrⁿ ′
00	5 45.0	5 45.9	5 29.3	0.0	0.0	6.0	2.4	12.0	4.7
01	5 45.3	5 46.2	5 29.5	0.1	0.0	6.1	2.4	12.1	4.7
02	5 45.5	5 46.4	5 29.8	0.2	0.1	6.2	2.4	12.2	4.8
03	5 45.8	5 46.7	5 30.0	0.3	0.1	6.3	2.5	12.3	4.8
04	5 46.0	5 46.9	5 30.2	0.4	0.2	6.4	2.5	12.4	4.9
05	5 46.3	5 47.2	5 30.5	0.5	0.2	6.5	2.5	12.5	4.9
06	5 46.5	5 47.4	5 30.7	0.6	0.2	6.6	2.6	12.6	4.9
07	5 46.8	5 47.7	5 31.0	0.7	0.3	6.7	2.6	12.7	5.0
08	5 47.0	5 48.0	5 31.2	0.8	0.3	6.8	2.7	12.8	5.0
09	5 47.3	5 48.2	5 31.4	0.9	0.4	6.9	2.7	12.9	5.1
10	5 47.5	5 48.5	5 31.7	1.0	0.4	7.0	2.7	13.0	5.1
11	5 47.8	5 48.7	5 31.9	1.1	0.4	7.1	2.8	13.1	5.1
12	5 48.0	5 49.0	5 32.1	1.2	0.5	7.2	2.8	13.2	5.2
13	5 48.3	5 49.2	5 32.4	1.3	0.5	7.3	2.9	13.3	5.2
14	5 48.5	5 49.5	5 32.6	1.4	0.5	7.4	2.9	13.4	5.2
15	5 48.8	5 49.7	5 32.9	1.5	0.6	7.5	2.9	13.5	5.3
16	5 49.0	5 50.0	5 33.1	1.6	0.6	7.6	3.0	13.6	5.3
17	5 49.3	5 50.2	5 33.3	1.7	0.7	7.7	3.0	13.7	5.4
18	5 49.5	5 50.5	5 33.6	1.8	0.7	7.8	3.1	13.8	5.4
19	5 49.8	5 50.7	5 33.8	1.9	0.7	7.9	3.1	13.9	5.4
20	5 50.0	5 51.0	5 34.1	2.0	0.8	8.0	3.1	14.0	5.5
21	5 50.3	5 51.2	5 34.3	2.1	0.8	8.1	3.2	14.1	5.5
22	5 50.5	5 51.5	5 34.5	2.2	0.9	8.2	3.2	14.2	5.6
23	5 50.8	5 51.7	5 34.8	2.3	0.9	8.3	3.3	14.3	5.6
24	5 51.0	5 52.0	5 35.0	2.4	0.9	8.4	3.3	14.4	5.6
25	5 51.3	5 52.2	5 35.2	2.5	1.0	8.5	3.3	14.5	5.7
26	5 51.5	5 52.5	5 35.5	2.6	1.0	8.6	3.4	14.6	5.7
27	5 51.8	5 52.7	5 35.7	2.7	1.1	8.7	3.4	14.7	5.8
28	5 52.0	5 53.0	5 36.0	2.8	1.1	8.8	3.4	14.8	5.8
29	5 52.3	5 53.2	5 36.2	2.9	1.1	8.9	3.5	14.9	5.8
30	5 52.5	5 53.5	5 36.4	3.0	1.2	9.0	3.5	15.0	5.9
31	5 52.8	5 53.7	5 36.7	3.1	1.2	9.1	3.6	15.1	5.9
32	5 53.0	5 54.0	5 36.9	3.2	1.3	9.2	3.6	15.2	6.0
33	5 53.3	5 54.2	5 37.2	3.3	1.3	9.3	3.6	15.3	6.0
34	5 53.5	5 54.5	5 37.4	3.4	1.3	9.4	3.7	15.4	6.0
35	5 53.8	5 54.7	5 37.6	3.5	1.4	9.5	3.7	15.5	6.1
36	5 54.0	5 55.0	5 37.9	3.6	1.4	9.6	3.8	15.6	6.1
37	5 54.3	5 55.2	5 38.1	3.7	1.4	9.7	3.8	15.7	6.1
38	5 54.5	5 55.5	5 38.4	3.8	1.5	9.8	3.8	15.8	6.2
39	5 54.8	5 55.7	5 38.6	3.9	1.5	9.9	3.9	15.9	6.2
40	5 55.0	5 56.0	5 38.8	4.0	1.6	10.0	3.9	16.0	6.3
41	5 55.3	5 56.2	5 39.1	4.1	1.6	10.1	4.0	16.1	6.3
42	5 55.5	5 56.5	5 39.3	4.2	1.6	10.2	4.0	16.2	6.3
43	5 55.8	5 56.7	5 39.5	4.3	1.7	10.3	4.0	16.3	6.4
44	5 56.0	5 57.0	5 39.8	4.4	1.7	10.4	4.1	16.4	6.4
45	5 56.3	5 57.2	5 40.0	4.5	1.8	10.5	4.1	16.5	6.5
46	5 56.5	5 57.5	5 40.3	4.6	1.8	10.6	4.2	16.6	6.5
47	5 56.8	5 57.7	5 40.5	4.7	1.8	10.7	4.2	16.7	6.5
48	5 57.0	5 58.0	5 40.7	4.8	1.9	10.8	4.2	16.8	6.6
49	5 57.3	5 58.2	5 41.0	4.9	1.9	10.9	4.3	16.9	6.6
50	5 57.5	5 58.5	5 41.2	5.0	2.0	11.0	4.3	17.0	6.7
51	5 57.8	5 58.7	5 41.5	5.1	2.0	11.1	4.3	17.1	6.7
52	5 58.0	5 59.0	5 41.7	5.2	2.0	11.2	4.4	17.2	6.7
53	5 58.3	5 59.2	5 41.9	5.3	2.1	11.3	4.4	17.3	6.8
54	5 58.5	5 59.5	5 42.2	5.4	2.1	11.4	4.5	17.4	6.8
55	5 58.8	5 59.7	5 42.4	5.5	2.2	11.5	4.5	17.5	6.9
56	5 59.0	6 00.0	5 42.6	5.6	2.2	11.6	4.5	17.6	6.9
57	5 59.3	6 00.2	5 42.9	5.7	2.2	11.7	4.6	17.7	6.9
58	5 59.5	6 00.5	5 43.1	5.8	2.3	11.8	4.6	17.8	7.0
59	5 59.8	6 00.7	5 43.4	5.9	2.3	11.9	4.7	17.9	7.0
60	6 00.0	6 01.0	5 43.6	6.0	2.4	12.0	4.7	18.0	7.1

24ᵐ	SUN PLANETS	ARIES	MOON	v or Corrⁿ d	v or Corrⁿ d	v or Corrⁿ d		25ᵐ	SUN PLANETS	ARIES	MOON	v or Corrⁿ d	v or Corrⁿ d	v or Corrⁿ d
s	° ′	° ′	° ′	′ ′	′ ′	′ ′		s	° ′	° ′	° ′	′ ′	′ ′	′ ′
00	6 00·0	6 01·0	5 43·6	0·0 0·0	6·0 2·5	12·0 4·9		00	6 15·0	6 16·0	5 57·9	0·0 0·0	6·0 2·6	12·0 5·1
01	6 00·3	6 01·2	5 43·8	0·1 0·0	6·1 2·5	12·1 4·9		01	6 15·3	6 16·3	5 58·2	0·1 0·0	6·1 2·6	12·1 5·1
02	6 00·5	6 01·5	5 44·1	0·2 0·1	6·2 2·5	12·2 5·0		02	6 15·5	6 16·5	5 58·4	0·2 0·1	6·2 2·6	12·2 5·2
03	6 00·8	6 01·7	5 44·3	0·3 0·1	6·3 2·6	12·3 5·0		03	6 15·8	6 16·8	5 58·6	0·3 0·1	6·3 2·7	12·3 5·2
04	6 01·0	6 02·0	5 44·6	0·4 0·2	6·4 2·6	12·4 5·1		04	6 16·0	6 17·0	5 58·9	0·4 0·2	6·4 2·7	12·4 5·3
05	6 01·3	6 02·2	5 44·8	0·5 0·2	6·5 2·7	12·5 5·1		05	6 16·3	6 17·3	5 59·1	0·5 0·2	6·5 2·8	12·5 5·3
06	6 01·5	6 02·5	5 45·0	0·6 0·2	6·6 2·7	12·6 5·1		06	6 16·5	6 17·5	5 59·3	0·6 0·3	6·6 2·8	12·6 5·4
07	6 01·8	6 02·7	5 45·3	0·7 0·3	6·7 2·7	12·7 5·2		07	6 16·8	6 17·8	5 59·6	0·7 0·3	6·7 2·8	12·7 5·4
08	6 02·0	6 03·0	5 45·5	0·8 0·3	6·8 2·8	12·8 5·2		08	6 17·0	6 18·0	5 59·8	0·8 0·3	6·8 2·9	12·8 5·4
09	6 02·3	6 03·2	5 45·7	0·9 0·4	6·9 2·8	12·9 5·3		09	6 17·3	6 18·3	6 00·1	0·9 0·4	6·9 2·9	12·9 5·5
10	6 02·5	6 03·5	5 46·0	1·0 0·4	7·0 2·9	13·0 5·3		10	6 17·5	6 18·5	6 00·3	1·0 0·4	7·0 3·0	13·0 5·5
11	6 02·8	6 03·7	5 46·2	1·1 0·4	7·1 2·9	13·1 5·3		11	6 17·8	6 18·8	6 00·5	1·1 0·5	7·1 3·0	13·1 5·6
12	6 03·0	6 04·0	5 46·5	1·2 0·5	7·2 2·9	13·2 5·4		12	6 18·0	6 19·0	6 00·8	1·2 0·5	7·2 3·1	13·2 5·6
13	6 03·3	6 04·2	5 46·7	1·3 0·5	7·3 3·0	13·3 5·4		13	6 18·3	6 19·3	6 01·0	1·3 0·6	7·3 3·1	13·3 5·7
14	6 03·5	6 04·5	5 46·9	1·4 0·6	7·4 3·0	13·4 5·5		14	6 18·5	6 19·5	6 01·3	1·4 0·6	7·4 3·1	13·4 5·7
15	6 03·8	6 04·7	5 47·2	1·5 0·6	7·5 3·1	13·5 5·5		15	6 18·8	6 19·8	6 01·5	1·5 0·6	7·5 3·2	13·5 5·7
16	6 04·0	6 05·0	5 47·4	1·6 0·7	7·6 3·1	13·6 5·6		16	6 19·0	6 20·0	6 01·7	1·6 0·7	7·6 3·2	13·6 5·8
17	6 04·3	6 05·2	5 47·7	1·7 0·7	7·7 3·1	13·7 5·6		17	6 19·3	6 20·3	6 02·0	1·7 0·7	7·7 3·3	13·7 5·8
18	6 04·5	6 05·5	5 47·9	1·8 0·7	7·8 3·2	13·8 5·6		18	6 19·5	6 20·5	6 02·2	1·8 0·8	7·8 3·3	13·8 5·9
19	6 04·8	6 05·7	5 48·1	1·9 0·8	7·9 3·2	13·9 5·7		19	6 19·8	6 20·8	6 02·5	1·9 0·8	7·9 3·4	13·9 5·9
20	6 05·0	6 06·0	5 48·4	2·0 0·8	8·0 3·3	14·0 5·7		20	6 20·0	6 21·0	6 02·7	2·0 0·9	8·0 3·4	14·0 6·0
21	6 05·3	6 06·3	5 48·6	2·1 0·9	8·1 3·3	14·1 5·8		21	6 20·3	6 21·3	6 02·9	2·1 0·9	8·1 3·4	14·1 6·0
22	6 05·5	6 06·5	5 48·8	2·2 0·9	8·2 3·3	14·2 5·8		22	6 20·5	6 21·5	6 03·2	2·2 0·9	8·2 3·5	14·2 6·0
23	6 05·8	6 06·8	5 49·1	2·3 0·9	8·3 3·4	14·3 5·8		23	6 20·8	6 21·8	6 03·4	2·3 1·0	8·3 3·5	14·3 6·1
24	6 06·0	6 07·0	5 49·3	2·4 1·0	8·4 3·4	14·4 5·9		24	6 21·0	6 22·0	6 03·6	2·4 1·0	8·4 3·6	14·4 6·1
25	6 06·3	6 07·3	5 49·6	2·5 1·0	8·5 3·5	14·5 5·9		25	6 21·3	6 22·3	6 03·9	2·5 1·1	8·5 3·6	14·5 6·2
26	6 06·5	6 07·5	5 49·8	2·6 1·1	8·6 3·5	14·6 6·0		26	6 21·5	6 22·5	6 04·1	2·6 1·1	8·6 3·7	14·6 6·2
27	6 06·8	6 07·8	5 50·0	2·7 1·1	8·7 3·6	14·7 6·0		27	6 21·8	6 22·8	6 04·4	2·7 1·1	8·7 3·7	14·7 6·2
28	6 07·0	6 08·0	5 50·3	2·8 1·1	8·8 3·6	14·8 6·0		28	6 22·0	6 23·0	6 04·6	2·8 1·2	8·8 3·7	14·8 6·3
29	6 07·3	6 08·3	5 50·5	2·9 1·2	8·9 3·6	14·9 6·1		29	6 22·3	6 23·3	6 04·8	2·9 1·2	8·9 3·8	14·9 6·3
30	6 07·5	6 08·5	5 50·8	3·0 1·2	9·0 3·7	15·0 6·1		30	6 22·5	6 23·5	6 05·1	3·0 1·3	9·0 3·8	15·0 6·4
31	6 07·8	6 08·8	5 51·0	3·1 1·3	9·1 3·7	15·1 6·2		31	6 22·8	6 23·8	6 05·3	3·1 1·3	9·1 3·9	15·1 6·4
32	6 08·0	6 09·0	5 51·2	3·2 1·3	9·2 3·8	15·2 6·2		32	6 23·0	6 24·0	6 05·6	3·2 1·4	9·2 3·9	15·2 6·5
33	6 08·3	6 09·3	5 51·5	3·3 1·3	9·3 3·8	15·3 6·2		33	6 23·3	6 24·3	6 05·8	3·3 1·4	9·3 4·0	15·3 6·5
34	6 08·5	6 09·5	5 51·7	3·4 1·4	9·4 3·8	15·4 6·3		34	6 23·5	6 24·5	6 06·0	3·4 1·4	9·4 4·0	15·4 6·5
35	6 08·8	6 09·8	5 52·0	3·5 1·4	9·5 3·9	15·5 6·3		35	6 23·8	6 24·8	6 06·3	3·5 1·5	9·5 4·0	15·5 6·6
36	6 09·0	6 10·0	5 52·2	3·6 1·5	9·6 3·9	15·6 6·4		36	6 24·0	6 25·1	6 06·5	3·6 1·5	9·6 4·1	15·6 6·6
37	6 09·3	6 10·3	5 52·4	3·7 1·5	9·7 4·0	15·7 6·4		37	6 24·3	6 25·3	6 06·7	3·7 1·6	9·7 4·1	15·7 6·7
38	6 09·5	6 10·5	5 52·7	3·8 1·6	9·8 4·0	15·8 6·5		38	6 24·5	6 25·6	6 07·0	3·8 1·6	9·8 4·2	15·8 6·7
39	6 09·8	6 10·8	5 52·9	3·9 1·6	9·9 4·0	15·9 6·5		39	6 24·8	6 25·8	6 07·2	3·9 1·7	9·9 4·2	15·9 6·8
40	6 10·0	6 11·0	5 53·1	4·0 1·6	10·0 4·1	16·0 6·5		40	6 25·0	6 26·1	6 07·5	4·0 1·7	10·0 4·3	16·0 6·8
41	6 10·3	6 11·3	5 53·4	4·1 1·7	10·1 4·1	16·1 6·6		41	6 25·3	6 26·3	6 07·7	4·1 1·7	10·1 4·3	16·1 6·9
42	6 10·5	6 11·5	5 53·6	4·2 1·7	10·2 4·2	16·2 6·6		42	6 25·5	6 26·6	6 07·9	4·2 1·8	10·2 4·3	16·2 6·9
43	6 10·8	6 11·8	5 53·9	4·3 1·8	10·3 4·2	16·3 6·7		43	6 25·8	6 26·8	6 08·2	4·3 1·8	10·3 4·4	16·3 6·9
44	6 11·0	6 12·0	5 54·1	4·4 1·8	10·4 4·2	16·4 6·7		44	6 26·0	6 27·1	6 08·4	4·4 1·9	10·4 4·4	16·4 7·0
45	6 11·3	6 12·3	5 54·3	4·5 1·8	10·5 4·3	16·5 6·7		45	6 26·3	6 27·3	6 08·7	4·5 1·9	10·5 4·5	16·5 7·0
46	6 11·5	6 12·5	5 54·6	4·6 1·9	10·6 4·3	16·6 6·8		46	6 26·5	6 27·6	6 08·9	4·6 2·0	10·6 4·5	16·6 7·1
47	6 11·8	6 12·8	5 54·8	4·7 1·9	10·7 4·4	16·7 6·8		47	6 26·8	6 27·8	6 09·1	4·7 2·0	10·7 4·5	16·7 7·1
48	6 12·0	6 13·0	5 55·1	4·8 2·0	10·8 4·4	16·8 6·9		48	6 27·0	6 28·1	6 09·4	4·8 2·0	10·8 4·6	16·8 7·1
49	6 12·3	6 13·3	5 55·3	4·9 2·0	10·9 4·5	16·9 6·9		49	6 27·3	6 28·3	6 09·6	4·9 2·1	10·9 4·6	16·9 7·2
50	6 12·5	6 13·5	5 55·5	5·0 2·0	11·0 4·5	17·0 6·9		50	6 27·5	6 28·6	6 09·8	5·0 2·1	11·0 4·7	17·0 7·2
51	6 12·8	6 13·8	5 55·8	5·1 2·1	11·1 4·5	17·1 7·0		51	6 27·8	6 28·8	6 10·1	5·1 2·2	11·1 4·7	17·1 7·3
52	6 13·0	6 14·0	5 56·0	5·2 2·1	11·2 4·6	17·2 7·0		52	6 28·0	6 29·1	6 10·3	5·2 2·2	11·2 4·8	17·2 7·3
53	6 13·3	6 14·3	5 56·2	5·3 2·2	11·3 4·6	17·3 7·1		53	6 28·3	6 29·3	6 10·6	5·3 2·3	11·3 4·8	17·3 7·4
54	6 13·5	6 14·5	5 56·5	5·4 2·2	11·4 4·7	17·4 7·1		54	6 28·5	6 29·6	6 10·8	5·4 2·3	11·4 4·8	17·4 7·4
55	6 13·8	6 14·8	5 56·7	5·5 2·2	11·5 4·7	17·5 7·1		55	6 28·8	6 29·8	6 11·0	5·5 2·3	11·5 4·9	17·5 7·4
56	6 14·0	6 15·0	5 57·0	5·6 2·3	11·6 4·7	17·6 7·2		56	6 29·0	6 30·1	6 11·3	5·6 2·4	11·6 4·9	17·6 7·5
57	6 14·3	6 15·3	5 57·2	5·7 2·3	11·7 4·8	17·7 7·2		57	6 29·3	6 30·3	6 11·5	5·7 2·4	11·7 5·0	17·7 7·5
58	6 14·5	6 15·5	5 57·4	5·8 2·4	11·8 4·8	17·8 7·3		58	6 29·5	6 30·6	6 11·8	5·8 2·5	11·8 5·0	17·8 7·6
59	6 14·8	6 15·8	5 57·7	5·9 2·4	11·9 4·9	17·9 7·3		59	6 29·8	6 30·8	6 12·0	5·9 2·5	11·9 5·1	17·9 7·6
60	6 15·0	6 16·0	5 57·9	6·0 2·5	12·0 4·9	18·0 7·4		60	6 30·0	6 31·1	6 12·2	6·0 2·6	12·0 5·1	18·0 7·7

52^m	SUN PLANETS	ARIES	MOON	v or d Corrⁿ		v or d Corrⁿ		v or d Corrⁿ	
s	° ′	° ′	° ′	′	′	′	′	′	′
00	13 00·0	13 02·1	12 24·5	0·0	0·0	6·0	5·3	12·0	10·5
01	13 00·3	13 02·4	12 24·7	0·1	0·1	6·1	5·3	12·1	10·6
02	13 00·5	13 02·6	12 24·9	0·2	0·2	6·2	5·4	12·2	10·7
03	13 00·8	13 02·9	12 25·2	0·3	0·3	6·3	5·5	12·3	10·8
04	13 01·0	13 03·1	12 25·4	0·4	0·4	6·4	5·6	12·4	10·9
05	13 01·3	13 03·4	12 25·7	0·5	0·4	6·5	5·7	12·5	10·9
06	13 01·5	13 03·6	12 25·9	0·6	0·5	6·6	5·8	12·6	11·0
07	13 01·8	13 03·9	12 26·1	0·7	0·6	6·7	5·9	12·7	11·1
08	13 02·0	13 04·1	12 26·4	0·8	0·7	6·8	6·0	12·8	11·2
09	13 02·3	13 04·4	12 26·6	0·9	0·8	6·9	6·0	12·9	11·3
10	13 02·5	13 04·6	12 26·9	1·0	0·9	7·0	6·1	13·0	11·4
11	13 02·8	13 04·9	12 27·1	1·1	1·0	7·1	6·2	13·1	11·5
12	13 03·0	13 05·1	12 27·3	1·2	1·1	7·2	6·3	13·2	11·6
13	13 03·3	13 05·4	12 27·6	1·3	1·1	7·3	6·4	13·3	11·6
14	13 03·5	13 05·6	12 27·8	1·4	1·2	7·4	6·5	13·4	11·7
15	13 03·8	13 05·9	12 28·0	1·5	1·3	7·5	6·6	13·5	11·8
16	13 04·0	13 06·1	12 28·3	1·6	1·4	7·6	6·7	13·6	11·9
17	13 04·3	13 06·4	12 28·5	1·7	1·5	7·7	6·7	13·7	12·0
18	13 04·5	13 06·6	12 28·8	1·8	1·6	7·8	6·8	13·8	12·1
19	13 04·8	13 06·9	12 29·0	1·9	1·7	7·9	6·9	13·9	12·2
20	13 05·0	13 07·1	12 29·2	2·0	1·8	8·0	7·0	14·0	12·3
21	13 05·3	13 07·4	12 29·5	2·1	1·8	8·1	7·1	14·1	12·3
22	13 05·5	13 07·7	12 29·7	2·2	1·9	8·2	7·2	14·2	12·4
23	13 05·8	13 07·9	12 30·0	2·3	2·0	8·3	7·3	14·3	12·5
24	13 06·0	13 08·2	12 30·2	2·4	2·1	8·4	7·4	14·4	12·6
25	13 06·3	13 08·4	12 30·4	2·5	2·2	8·5	7·4	14·5	12·7
26	13 06·5	13 08·7	12 30·7	2·6	2·3	8·6	7·5	14·6	12·8
27	13 06·8	13 08·9	12 30·9	2·7	2·4	8·7	7·6	14·7	12·9
28	13 07·0	13 09·2	12 31·1	2·8	2·5	8·8	7·7	14·8	13·0
29	13 07·3	13 09·4	12 31·4	2·9	2·5	8·9	7·8	14·9	13·0
30	13 07·5	13 09·7	12 31·6	3·0	2·6	9·0	7·9	15·0	13·1
31	13 07·8	13 09·9	12 31·9	3·1	2·7	9·1	8·0	15·1	13·2
32	13 08·0	13 10·2	12 32·1	3·2	2·8	9·2	8·1	15·2	13·3
33	13 08·3	13 10·4	12 32·3	3·3	2·9	9·3	8·1	15·3	13·4
34	13 08·5	13 10·7	12 32·6	3·4	3·0	9·4	8·2	15·4	13·5
35	13 08·8	13 10·9	12 32·8	3·5	3·1	9·5	8·3	15·5	13·6
36	13 09·0	13 11·2	12 33·1	3·6	3·2	9·6	8·4	15·6	13·7
37	13 09·3	13 11·4	12 33·3	3·7	3·2	9·7	8·5	15·7	13·7
38	13 09·5	13 11·7	12 33·5	3·8	3·3	9·8	8·6	15·8	13·8
39	13 09·8	13 11·9	12 33·8	3·9	3·4	9·9	8·7	15·9	13·9
40	13 10·0	13 12·2	12 34·0	4·0	3·5	10·0	8·8	16·0	14·0
41	13 10·3	13 12·4	12 34·2	4·1	3·6	10·1	8·8	16·1	14·1
42	13 10·5	13 12·7	12 34·5	4·2	3·7	10·2	8·9	16·2	14·2
43	13 10·8	13 12·9	12 34·7	4·3	3·8	10·3	9·0	16·3	14·3
44	13 11·0	13 13·2	12 35·0	4·4	3·9	10·4	9·1	16·4	14·4
45	13 11·3	13 13·4	12 35·2	4·5	3·9	10·5	9·2	16·5	14·4
46	13 11·5	13 13·7	12 35·4	4·6	4·0	10·6	9·3	16·6	14·5
47	13 11·8	13 13·9	12 35·7	4·7	4·1	10·7	9·4	16·7	14·6
48	13 12·0	13 14·2	12 35·9	4·8	4·2	10·8	9·5	16·8	14·7
49	13 12·3	13 14·4	12 36·2	4·9	4·3	10·9	9·5	16·9	14·8
50	13 12·5	13 14·7	12 36·4	5·0	4·4	11·0	9·6	17·0	14·9
51	13 12·8	13 14·9	12 36·6	5·1	4·5	11·1	9·7	17·1	15·0
52	13 13·0	13 15·2	12 36·9	5·2	4·6	11·2	9·8	17·2	15·1
53	13 13·3	13 15·4	12 37·1	5·3	4·6	11·3	9·9	17·3	15·1
54	13 13·5	13 15·7	12 37·4	5·4	4·7	11·4	10·0	17·4	15·2
55	13 13·8	13 15·9	12 37·6	5·5	4·8	11·5	10·1	17·5	15·3
56	13 14·0	13 16·2	12 37·8	5·6	4·9	11·6	10·2	17·6	15·4
57	13 14·3	13 16·4	12 38·1	5·7	5·0	11·7	10·2	17·7	15·5
58	13 14·5	13 16·7	12 38·3	5·8	5·1	11·8	10·3	17·8	15·6
59	13 14·8	13 16·9	12 38·5	5·9	5·2	11·9	10·4	17·9	15·7
60	13 15·0	13 17·2	12 38·8	6·0	5·3	12·0	10·5	18·0	15·8

53^m	SUN PLANETS	ARIES	MOON	v or d Corrⁿ		v or d Corrⁿ		v or d Corrⁿ	
s	° ′	° ′	° ′	′	′	′	′	′	′
00	13 15·0	13 17·2	12 38·8	0·0	0·0	6·0	5·4	12·0	10·7
01	13 15·3	13 17·4	12 39·0	0·1	0·1	6·1	5·4	12·1	10·8
02	13 15·5	13 17·7	12 39·3	0·2	0·2	6·2	5·5	12·2	10·9
03	13 15·8	13 17·9	12 39·5	0·3	0·3	6·3	5·6	12·3	11·0
04	13 16·0	13 18·2	12 39·7	0·4	0·4	6·4	5·7	12·4	11·1
05	13 16·3	13 18·4	12 40·0	0·5	0·4	6·5	5·8	12·5	11·1
06	13 16·5	13 18·7	12 40·2	0·6	0·5	6·6	5·9	12·6	11·2
07	13 16·8	13 18·9	12 40·5	0·7	0·6	6·7	6·0	12·7	11·3
08	13 17·0	13 19·2	12 40·7	0·8	0·7	6·8	6·1	12·8	11·4
09	13 17·3	13 19·4	12 40·9	0·9	0·8	6·9	6·2	12·9	11·5
10	13 17·5	13 19·7	12 41·2	1·0	0·9	7·0	6·2	13·0	11·6
11	13 17·8	13 19·9	12 41·4	1·1	1·0	7·1	6·3	13·1	11·7
12	13 18·0	13 20·2	12 41·6	1·2	1·1	7·2	6·4	13·2	11·8
13	13 18·3	13 20·4	12 41·9	1·3	1·2	7·3	6·5	13·3	11·9
14	13 18·5	13 20·7	12 42·1	1·4	1·2	7·4	6·6	13·4	11·9
15	13 18·8	13 20·9	12 42·4	1·5	1·3	7·5	6·7	13·5	12·0
16	13 19·0	13 21·2	12 42·6	1·6	1·4	7·6	6·8	13·6	12·1
17	13 19·3	13 21·4	12 42·8	1·7	1·5	7·7	6·9	13·7	12·2
18	13 19·5	13 21·7	12 43·1	1·8	1·6	7·8	7·0	13·8	12·3
19	13 19·8	13 21·9	12 43·3	1·9	1·7	7·9	7·0	13·9	12·4
20	13 20·0	13 22·2	12 43·6	2·0	1·8	8·0	7·1	14·0	12·5
21	13 20·3	13 22·4	12 43·8	2·1	1·9	8·1	7·2	14·1	12·6
22	13 20·5	13 22·7	12 44·0	2·2	2·0	8·2	7·3	14·2	12·7
23	13 20·8	13 22·9	12 44·3	2·3	2·1	8·3	7·4	14·3	12·8
24	13 21·0	13 23·2	12 44·5	2·4	2·1	8·4	7·5	14·4	12·8
25	13 21·3	13 23·4	12 44·7	2·5	2·2	8·5	7·6	14·5	12·9
26	13 21·5	13 23·7	12 45·0	2·6	2·3	8·6	7·7	14·6	13·0
27	13 21·8	13 23·9	12 45·2	2·7	2·4	8·7	7·8	14·7	13·1
28	13 22·0	13 24·2	12 45·5	2·8	2·5	8·8	7·8	14·8	13·2
29	13 22·3	13 24·4	12 45·7	2·9	2·6	8·9	7·9	14·9	13·3
30	13 22·5	13 24·7	12 45·9	3·0	2·7	9·0	8·0	15·0	13·4
31	13 22·8	13 24·9	12 46·2	3·1	2·8	9·1	8·1	15·1	13·5
32	13 23·0	13 25·2	12 46·4	3·2	2·9	9·2	8·2	15·2	13·6
33	13 23·3	13 25·4	12 46·7	3·3	2·9	9·3	8·3	15·3	13·6
34	13 23·5	13 25·7	12 46·9	3·4	3·0	9·4	8·4	15·4	13·7
35	13 23·8	13 26·0	12 47·1	3·5	3·1	9·5	8·5	15·5	13·8
36	13 24·0	13 26·2	12 47·4	3·6	3·2	9·6	8·6	15·6	13·9
37	13 24·3	13 26·5	12 47·6	3·7	3·3	9·7	8·6	15·7	14·0
38	13 24·5	13 26·7	12 47·9	3·8	3·4	9·8	8·7	15·8	14·1
39	13 24·8	13 27·0	12 48·1	3·9	3·5	9·9	8·8	15·9	14·2
40	13 25·0	13 27·2	12 48·3	4·0	3·6	10·0	8·9	16·0	14·3
41	13 25·3	13 27·5	12 48·6	4·1	3·7	10·1	9·0	16·1	14·4
42	13 25·5	13 27·7	12 48·8	4·2	3·7	10·2	9·1	16·2	14·4
43	13 25·8	13 28·0	12 49·0	4·3	3·8	10·3	9·2	16·3	14·5
44	13 26·0	13 28·2	12 49·3	4·4	3·9	10·4	9·3	16·4	14·6
45	13 26·3	13 28·5	12 49·5	4·5	4·0	10·5	9·4	16·5	14·7
46	13 26·5	13 28·7	12 49·8	4·6	4·1	10·6	9·5	16·6	14·8
47	13 26·8	13 29·0	12 50·0	4·7	4·2	10·7	9·5	16·7	14·9
48	13 27·0	13 29·2	12 50·2	4·8	4·3	10·8	9·6	16·8	15·0
49	13 27·3	13 29·5	12 50·5	4·9	4·4	10·9	9·7	16·9	15·1
50	13 27·5	13 29·7	12 50·7	5·0	4·5	11·0	9·8	17·0	15·2
51	13 27·8	13 30·0	12 51·0	5·1	4·5	11·1	9·9	17·1	15·2
52	13 28·0	13 30·2	12 51·2	5·2	4·6	11·2	10·0	17·2	15·3
53	13 28·3	13 30·5	12 51·4	5·3	4·7	11·3	10·1	17·3	15·4
54	13 28·5	13 30·7	12 51·7	5·4	4·8	11·4	10·2	17·4	15·5
55	13 28·8	13 31·0	12 51·9	5·5	4·9	11·5	10·3	17·5	15·6
56	13 29·0	13 31·2	12 52·1	5·6	5·0	11·6	10·3	17·6	15·7
57	13 29·3	13 31·5	12 52·4	5·7	5·1	11·7	10·4	17·7	15·8
58	13 29·5	13 31·7	12 52·6	5·8	5·2	11·8	10·5	17·8	15·9
59	13 29·8	13 32·0	12 52·9	5·9	5·3	11·9	10·6	17·9	16·0
60	13 30·0	13 32·2	12 53·1	6·0	5·4	12·0	10·7	18·0	16·1

56ᵐ

56ᵐ s	SUN PLANETS ° ′	ARIES ° ′	MOON ° ′	v or d	Corrⁿ ′	v or d	Corrⁿ ′	v or d	Corrⁿ ′
00	14 00·0	14 02·3	13 21·7	0·0	0·0	6·0	5·7	12·0	11·3
01	14 00·3	14 02·6	13 22·0	0·1	0·1	6·1	5·7	12·1	11·4
02	14 00·5	14 02·8	13 22·2	0·2	0·2	6·2	5·8	12·2	11·5
03	14 00·8	14 03·1	13 22·4	0·3	0·3	6·3	5·9	12·3	11·6
04	14 01·0	14 03·3	13 22·7	0·4	0·4	6·4	6·0	12·4	11·7
05	14 01·3	14 03·6	13 22·9	0·5	0·5	6·5	6·1	12·5	11·8
06	14 01·5	14 03·8	13 23·2	0·6	0·6	6·6	6·2	12·6	11·9
07	14 01·8	14 04·1	13 23·4	0·7	0·7	6·7	6·3	12·7	12·0
08	14 02·0	14 04·3	13 23·6	0·8	0·8	6·8	6·4	12·8	12·1
09	14 02·3	14 04·6	13 23·9	0·9	0·8	6·9	6·5	12·9	12·1
10	14 02·5	14 04·8	13 24·1	1·0	0·9	7·0	6·6	13·0	12·2
11	14 02·8	14 05·1	13 24·4	1·1	1·0	7·1	6·7	13·1	12·3
12	14 03·0	14 05·3	13 24·6	1·2	1·1	7·2	6·8	13·2	12·4
13	14 03·3	14 05·6	13 24·8	1·3	1·2	7·3	6·9	13·3	12·5
14	14 03·5	14 05·8	13 25·1	1·4	1·3	7·4	7·0	13·4	12·6
15	14 03·8	14 06·1	13 25·3	1·5	1·4	7·5	7·1	13·5	12·7
16	14 04·0	14 06·3	13 25·6	1·6	1·5	7·6	7·2	13·6	12·8
17	14 04·3	14 06·6	13 25·8	1·7	1·6	7·7	7·3	13·7	12·9
18	14 04·5	14 06·8	13 26·0	1·8	1·7	7·8	7·3	13·8	13·0
19	14 04·8	14 07·1	13 26·3	1·9	1·8	7·9	7·4	13·9	13·1
20	14 05·0	14 07·3	13 26·5	2·0	1·9	8·0	7·5	14·0	13·2
21	14 05·3	14 07·6	13 26·7	2·1	2·0	8·1	7·6	14·1	13·3
22	14 05·5	14 07·8	13 27·0	2·2	2·1	8·2	7·7	14·2	13·4
23	14 05·8	14 08·1	13 27·2	2·3	2·2	8·3	7·8	14·3	13·5
24	14 06·0	14 08·3	13 27·5	2·4	2·3	8·4	7·9	14·4	13·6
25	14 06·3	14 08·6	13 27·7	2·5	2·4	8·5	8·0	14·5	13·7
26	14 06·5	14 08·8	13 27·9	2·6	2·4	8·6	8·1	14·6	13·7
27	14 06·8	14 09·1	13 28·2	2·7	2·5	8·7	8·2	14·7	13·8
28	14 07·0	14 09·3	13 28·4	2·8	2·6	8·8	8·3	14·8	13·9
29	14 07·3	14 09·6	13 28·7	2·9	2·7	8·9	8·4	14·9	14·0
30	14 07·5	14 09·8	13 28·9	3·0	2·8	9·0	8·5	15·0	14·1
31	14 07·8	14 10·1	13 29·1	3·1	2·9	9·1	8·6	15·1	14·2
32	14 08·0	14 10·3	13 29·4	3·2	3·0	9·2	8·7	15·2	14·3
33	14 08·3	14 10·6	13 29·6	3·3	3·1	9·3	8·8	15·3	14·4
34	14 08·5	14 10·8	13 29·8	3·4	3·2	9·4	8·9	15·4	14·5
35	14 08·8	14 11·1	13 30·1	3·5	3·3	9·5	8·9	15·5	14·6
36	14 09·0	14 11·3	13 30·3	3·6	3·4	9·6	9·0	15·6	14·7
37	14 09·3	14 11·6	13 30·6	3·7	3·5	9·7	9·1	15·7	14·8
38	14 09·5	14 11·8	13 30·8	3·8	3·6	9·8	9·2	15·8	14·9
39	14 09·8	14 12·1	13 31·0	3·9	3·7	9·9	9·3	15·9	15·0
40	14 10·0	14 12·3	13 31·3	4·0	3·8	10·0	9·4	16·0	15·1
41	14 10·3	14 12·6	13 31·5	4·1	3·9	10·1	9·5	16·1	15·2
42	14 10·5	14 12·8	13 31·8	4·2	4·0	10·2	9·6	16·2	15·3
43	14 10·8	14 13·1	13 32·0	4·3	4·0	10·3	9·7	16·3	15·3
44	14 11·0	14 13·3	13 32·2	4·4	4·1	10·4	9·8	16·4	15·4
45	14 11·3	14 13·6	13 32·5	4·5	4·2	10·5	9·9	16·5	15·5
46	14 11·5	14 13·8	13 32·7	4·6	4·3	10·6	10·0	16·6	15·6
47	14 11·8	14 14·1	13 33·0	4·7	4·4	10·7	10·1	16·7	15·7
48	14 12·0	14 14·3	13 33·2	4·8	4·5	10·8	10·2	16·8	15·8
49	14 12·3	14 14·6	13 33·4	4·9	4·6	10·9	10·3	16·9	15·9
50	14 12·5	14 14·8	13 33·7	5·0	4·7	11·0	10·4	17·0	16·0
51	14 12·8	14 15·1	13 33·9	5·1	4·8	11·1	10·5	17·1	16·1
52	14 13·0	14 15·3	13 34·1	5·2	4·9	11·2	10·5	17·2	16·2
53	14 13·3	14 15·6	13 34·4	5·3	5·0	11·3	10·6	17·3	16·3
54	14 13·5	14 15·8	13 34·6	5·4	5·1	11·4	10·7	17·4	16·4
55	14 13·8	14 16·1	13 34·9	5·5	5·2	11·5	10·8	17·5	16·5
56	14 14·0	14 16·3	13 35·1	5·6	5·3	11·6	10·9	17·6	16·6
57	14 14·3	14 16·6	13 35·3	5·7	5·4	11·7	11·0	17·7	16·7
58	14 14·5	14 16·8	13 35·6	5·8	5·5	11·8	11·1	17·8	16·8
59	14 14·8	14 17·1	13 35·8	5·9	5·6	11·9	11·2	17·9	16·9
60	14 15·0	14 17·3	13 36·1	6·0	5·7	12·0	11·3	18·0	17·0

57ᵐ

57ᵐ s	SUN PLANETS ° ′	ARIES ° ′	MOON ° ′	v or d	Corrⁿ ′	v or d	Corrⁿ ′	v or d	Corrⁿ ′
00	14 15·0	14 17·3	13 36·1	0·0	0·0	6·0	5·8	12·0	11·5
01	14 15·3	14 17·6	13 36·3	0·1	0·1	6·1	5·8	12·1	11·6
02	14 15·5	14 17·8	13 36·5	0·2	0·2	6·2	5·9	12·2	11·7
03	14 15·8	14 18·1	13 36·8	0·3	0·3	6·3	6·0	12·3	11·8
04	14 16·0	14 18·3	13 37·0	0·4	0·4	6·4	6·1	12·4	11·9
05	14 16·3	14 18·6	13 37·2	0·5	0·5	6·5	6·2	12·5	12·0
06	14 16·5	14 18·8	13 37·5	0·6	0·6	6·6	6·3	12·6	12·1
07	14 16·8	14 19·1	13 37·7	0·7	0·7	6·7	6·4	12·7	12·2
08	14 17·0	14 19·3	13 38·0	0·8	0·8	6·8	6·5	12·8	12·3
09	14 17·3	14 19·6	13 38·2	0·9	0·9	6·9	6·6	12·9	12·4
10	14 17·5	14 19·8	13 38·4	1·0	1·0	7·0	6·7	13·0	12·5
11	14 17·8	14 20·1	13 38·7	1·1	1·1	7·1	6·8	13·1	12·6
12	14 18·0	14 20·3	13 38·9	1·2	1·2	7·2	6·9	13·2	12·7
13	14 18·3	14 20·6	13 39·2	1·3	1·2	7·3	7·0	13·3	12·7
14	14 18·5	14 20·9	13 39·4	1·4	1·3	7·4	7·1	13·4	12·8
15	14 18·8	14 21·1	13 39·6	1·5	1·4	7·5	7·2	13·5	12·9
16	14 19·0	14 21·4	13 39·9	1·6	1·5	7·6	7·3	13·6	13·0
17	14 19·3	14 21·6	13 40·1	1·7	1·6	7·7	7·4	13·7	13·1
18	14 19·5	14 21·9	13 40·3	1·8	1·7	7·8	7·5	13·8	13·2
19	14 19·8	14 22·1	13 40·6	1·9	1·8	7·9	7·6	13·9	13·3
20	14 20·0	14 22·4	13 40·8	2·0	1·9	8·0	7·7	14·0	13·4
21	14 20·3	14 22·6	13 41·1	2·1	2·0	8·1	7·8	14·1	13·5
22	14 20·5	14 22·9	13 41·3	2·2	2·1	8·2	7·9	14·2	13·6
23	14 20·8	14 23·1	13 41·5	2·3	2·2	8·3	8·0	14·3	13·7
24	14 21·0	14 23·4	13 41·8	2·4	2·3	8·4	8·1	14·4	13·8
25	14 21·3	14 23·6	13 42·0	2·5	2·4	8·5	8·1	14·5	13·9
26	14 21·5	14 23·9	13 42·3	2·6	2·5	8·6	8·2	14·6	14·0
27	14 21·8	14 24·1	13 42·5	2·7	2·6	8·7	8·3	14·7	14·1
28	14 22·0	14 24·4	13 42·7	2·8	2·7	8·8	8·4	14·8	14·2
29	14 22·3	14 24·6	13 43·0	2·9	2·8	8·9	8·5	14·9	14·3
30	14 22·5	14 24·9	13 43·2	3·0	2·9	9·0	8·6	15·0	14·4
31	14 22·8	14 25·1	13 43·4	3·1	3·0	9·1	8·7	15·1	14·5
32	14 23·0	14 25·4	13 43·7	3·2	3·1	9·2	8·8	15·2	14·6
33	14 23·3	14 25·6	13 43·9	3·3	3·2	9·3	8·9	15·3	14·7
34	14 23·5	14 25·9	13 44·2	3·4	3·3	9·4	9·0	15·4	14·8
35	14 23·8	14 26·1	13 44·4	3·5	3·4	9·5	9·1	15·5	14·9
36	14 24·0	14 26·4	13 44·6	3·6	3·5	9·6	9·2	15·6	15·0
37	14 24·3	14 26·6	13 44·9	3·7	3·5	9·7	9·3	15·7	15·0
38	14 24·5	14 26·9	13 45·1	3·8	3·6	9·8	9·4	15·8	15·1
39	14 24·8	14 27·1	13 45·4	3·9	3·7	9·9	9·5	15·9	15·2
40	14 25·0	14 27·4	13 45·6	4·0	3·8	10·0	9·6	16·0	15·3
41	14 25·3	14 27·6	13 45·8	4·1	3·9	10·1	9·7	16·1	15·4
42	14 25·5	14 27·9	13 46·1	4·2	4·0	10·2	9·8	16·2	15·5
43	14 25·8	14 28·1	13 46·3	4·3	4·1	10·3	9·9	16·3	15·6
44	14 26·0	14 28·4	13 46·5	4·4	4·2	10·4	10·0	16·4	15·7
45	14 26·3	14 28·6	13 46·8	4·5	4·3	10·5	10·1	16·5	15·8
46	14 26·5	14 28·9	13 47·0	4·6	4·4	10·6	10·2	16·6	15·9
47	14 26·8	14 29·1	13 47·3	4·7	4·5	10·7	10·3	16·7	16·0
48	14 27·0	14 29·4	13 47·5	4·8	4·6	10·8	10·4	16·8	16·1
49	14 27·3	14 29·6	13 47·7	4·9	4·7	10·9	10·4	16·9	16·2
50	14 27·5	14 29·9	13 48·0	5·0	4·8	11·0	10·5	17·0	16·3
51	14 27·8	14 30·1	13 48·2	5·1	4·9	11·1	10·6	17·1	16·4
52	14 28·0	14 30·4	13 48·5	5·2	5·0	11·2	10·7	17·2	16·5
53	14 28·3	14 30·6	13 48·7	5·3	5·1	11·3	10·8	17·3	16·6
54	14 28·5	14 30·9	13 48·9	5·4	5·2	11·4	10·9	17·4	16·7
55	14 28·8	14 31·1	13 49·2	5·5	5·3	11·5	11·0	17·5	16·8
56	14 29·0	14 31·4	13 49·4	5·6	5·4	11·6	11·1	17·6	16·9
57	14 29·3	14 31·6	13 49·7	5·7	5·5	11·7	11·2	17·7	17·0
58	14 29·5	14 31·9	13 49·9	5·8	5·6	11·8	11·3	17·8	17·1
59	14 29·8	14 32·1	13 50·1	5·9	5·7	11·9	11·4	17·9	17·2
60	14 30·0	14 32·4	13 50·4	6·0	5·8	12·0	11·5	18·0	17·3

App. Alt.	35°–39° Corrⁿ	40°–44° Corrⁿ	45°–49° Corrⁿ	50°–54° Corrⁿ	55°–59° Corrⁿ	60°–64° Corrⁿ	65°–69° Corrⁿ	70°–74° Corrⁿ	75°–79° Corrⁿ	80°–84° Corrⁿ	85°–89° Corrⁿ	App. Alt.
00	35 56.5	40 53.7	45 50.5	50 46.9	55 43.1	60 38.9	65 34.6	70 30.1	75 25.3	80 20.5	85 15.6	00
10	56.4	53.6	50.4	46.8	42.9	38.8	34.4	29.9	25.2	20.4	15.5	10
20	56.3	53.5	50.2	46.7	42.8	38.7	34.3	29.7	25.0	20.2	15.3	20
30	56.2	53.4	50.1	46.5	42.7	38.5	34.1	29.6	24.9	20.0	15.1	30
40	56.2	53.3	50.0	46.4	42.5	38.4	34.0	29.4	24.7	19.9	15.0	40
50	56.1	53.2	49.9	46.3	42.4	38.2	33.8	29.3	24.5	19.7	14.8	50
00	36 56.0	41 53.1	46 49.8	51 46.2	56 42.3	61 38.1	66 33.7	71 29.1	76 24.4	81 19.6	86 14.6	00
10	55.9	53.0	49.7	46.0	42.1	37.9	33.5	29.0	24.2	19.4	14.5	10
20	55.8	52.8	49.5	45.9	42.0	37.8	33.4	28.8	24.1	19.2	14.3	20
30	55.7	52.7	49.4	45.8	41.8	37.7	33.2	28.7	23.9	19.1	14.1	30
40	55.6	52.6	49.3	45.7	41.7	37.5	33.1	28.5	23.8	18.9	14.0	40
50	55.5	52.5	49.2	45.5	41.6	37.4	32.9	28.3	23.6	18.7	13.8	50
00	37 55.4	42 52.4	47 49.1	52 45.4	57 41.4	62 37.2	67 32.8	72 28.2	77 23.4	82 18.6	87 13.7	00
10	55.3	52.3	49.0	45.3	41.3	37.1	32.6	28.0	23.3	18.4	13.5	10
20	55.2	52.2	48.8	45.2	41.2	36.9	32.5	27.9	23.1	18.2	13.3	20
30	55.1	52.1	48.7	45.0	41.0	36.8	32.3	27.7	22.9	18.1	13.2	30
40	55.0	52.0	48.6	44.9	40.9	36.6	32.2	27.6	22.8	17.9	13.0	40
50	55.0	51.9	48.5	44.8	40.8	36.5	32.0	27.4	22.6	17.8	12.8	50
00	38 54.9	43 51.8	48 48.4	53 44.6	58 40.6	63 36.4	68 31.9	73 27.2	78 22.5	83 17.6	88 12.7	00
10	54.8	51.7	48.2	44.5	40.5	36.2	31.7	27.1	22.3	17.4	12.5	10
20	54.7	51.6	48.1	44.4	40.3	36.1	31.6	26.9	22.1	17.3	12.3	20
30	54.6	51.5	48.0	44.2	40.2	35.9	31.4	26.8	22.0	17.1	12.2	30
40	54.5	51.4	47.9	44.1	40.1	35.8	31.3	26.6	21.8	16.9	12.0	40
50	54.4	51.2	47.8	44.0	39.9	35.6	31.1	26.5	21.7	16.8	11.8	50
00	39 54.3	44 51.1	49 47.6	54 43.9	59 39.8	64 35.5	69 31.0	74 26.3	79 21.5	84 16.6	89 11.7	00
10	54.2	51.0	47.5	43.7	39.6	35.3	30.8	26.1	21.3	16.5	11.5	10
20	54.1	50.9	47.4	43.6	39.5	35.2	30.7	26.0	21.2	16.3	11.4	20
30	54.0	50.8	47.3	43.5	39.4	35.0	30.5	25.8	21.0	16.1	11.2	30
40	53.9	50.7	47.2	43.3	39.2	34.9	30.4	25.7	20.9	16.0	11.0	40
50	53.8	50.6	47.0	43.2	39.1	34.7	30.2	25.5	20.7	15.8	10.9	50

H.P.	L U	L U	L U	L U	L U	L U	L U	L U
54.0	1.1 1.7	1.3 1.9	1.5 2.1	1.7 2.4	2.0 2.6	2.3 2.9	2.6 3.2	2.9 3.5
54.3	1.4 1.8	1.6 2.0	1.8 2.2	2.0 2.5	2.3 2.7	2.5 3.0	2.8 3.2	3.0 3.5
54.6	1.7 2.0	1.9 2.2	2.1 2.4	2.3 2.6	2.5 2.8	2.7 3.0	3.0 3.3	3.2 3.5
54.9	2.0 2.2	2.2 2.3	2.3 2.5	2.5 2.7	2.7 2.9	2.9 3.1	3.2 3.3	3.4 3.5
55.2	2.3 2.3	2.5 2.4	2.6 2.6	2.8 2.8	3.0 2.9	3.2 3.1	3.4 3.3	3.6 3.5
55.5	2.7 2.5	2.8 2.6	2.9 2.7	3.1 2.9	3.2 3.0	3.4 3.2	3.6 3.4	3.7 3.5
55.8	3.0 2.6	3.1 2.7	3.2 2.8	3.3 3.0	3.5 3.1	3.6 3.3	3.8 3.4	3.9 3.6
56.1	3.3 2.8	3.4 2.9	3.5 3.0	3.6 3.1	3.7 3.2	3.8 3.3	4.0 3.4	4.1 3.6
56.4	3.6 2.9	3.7 3.0	3.8 3.1	3.9 3.2	3.9 3.3	4.0 3.4	4.1 3.5	4.3 3.6
56.7	3.9 3.1	4.0 3.1	4.1 3.2	4.1 3.3	4.2 3.3	4.3 3.4	4.3 3.5	4.4 3.6
57.0	4.3 3.2	4.3 3.3	4.3 3.3	4.4 3.4	4.4 3.4	4.5 3.5	4.5 3.5	4.6 3.6
57.3	4.6 3.4	4.6 3.4	4.6 3.4	4.6 3.5	4.7 3.5	4.7 3.5	4.7 3.6	4.8 3.6
57.6	4.9 3.6	4.9 3.6	4.9 3.6	4.9 3.6	4.9 3.6	4.9 3.6	4.9 3.6	4.9 3.6
57.9	5.2 3.7	5.2 3.7	5.2 3.7	5.2 3.7	5.2 3.7	5.1 3.6	5.1 3.6	5.1 3.6
58.2	5.5 3.9	5.5 3.8	5.5 3.8	5.4 3.8	5.4 3.7	5.4 3.7	5.3 3.7	5.3 3.6
58.5	5.9 4.0	5.8 4.0	5.8 3.9	5.7 3.9	5.6 3.8	5.6 3.8	5.5 3.7	5.5 3.6
58.8	6.2 4.2	6.1 4.1	6.0 4.1	6.0 4.0	5.9 3.9	5.8 3.8	5.7 3.7	5.6 3.6
59.1	6.5 4.3	6.4 4.3	6.3 4.2	6.2 4.1	6.1 4.0	6.0 3.9	5.9 3.8	5.8 3.6
59.4	6.8 4.5	6.7 4.4	6.6 4.3	6.5 4.2	6.4 4.1	6.2 3.9	6.1 3.8	6.0 3.7
59.7	7.1 4.6	7.0 4.5	6.9 4.4	6.8 4.3	6.6 4.1	6.5 4.0	6.3 3.8	6.2 3.7
60.0	7.5 4.8	7.3 4.7	7.2 4.5	7.0 4.4	6.9 4.2	6.7 4.0	6.5 3.9	6.3 3.7
60.3	7.8 5.0	7.6 4.8	7.5 4.7	7.3 4.5	7.1 4.3	6.9 4.1	6.7 3.9	6.5 3.7
60.6	8.1 5.1	7.9 5.0	7.7 4.8	7.6 4.6	7.3 4.4	7.1 4.2	6.9 3.9	6.7 3.7
60.9	8.4 5.3	8.2 5.1	8.0 4.9	7.8 4.7	7.6 4.5	7.3 4.2	7.1 4.0	6.8 3.7
61.2	8.7 5.4	8.5 5.2	8.3 5.0	8.1 4.8	7.8 4.5	7.6 4.3	7.3 4.0	7.0 3.7
61.5	9.1 5.6	8.8 5.4	8.6 5.1	8.3 4.9	8.1 4.6	7.8 4.3	7.5 4.0	7.2 3.7

MOON CORRECTION TABLE

The correction is in two parts; the first correction is taken from the upper part of the table with argument apparent altitude, and the second from the lower part, with argument H.P., in the same column as that from which the first correction was taken. Separate corrections are given in the lower part for lower (L) and upper (U) limbs. All corrections are to be **added** to apparent altitude, *but 30' is to be subtracted from the altitude of the upper limb.*

For corrections for pressure and temperature see page A4.

For bubble sextant observations ignore dip, take the mean of upper and lower limb corrections and subtract 15' from the altitude.

App. Alt. = Apparent altitude = Sextant altitude corrected for index error and dip.

TABLE FROM P. xxxiv

xxxv

N. Lat. {LHA greater than 180°....... Zn=Z / LHA less than 180°.......... Zn=360−Z}

DECLINATION (15°-29°) SAME NAME AS LATITUDE

LHA	15° Hc	15° d	15° Z	16° Hc	16° d	16° Z	17° Hc	17° d	17° Z	18° Hc	18° d	18° Z	19° Hc	19° d	19° Z
	° ′	′	°	° ′	′	°	° ′	′	°	° ′	′	°	° ′	′	°
0	63 00	+60	180	64 00	+60	180	65 00	+60	180	66 00	+60	180	67 00	+60	180
1	62 59	60	178	63 59	60	178	64 59	60	178	65 59	60	178	66 59	60	178
2	62 57	60	176	63 57	60	176	64 57	59	176	65 56	60	175	66 56	60	175
3	62 53	59	174	63 52	60	173	64 52	60	173	65 52	60	173	66 52	59	173
4	62 47	59	172	63 46	60	171	64 46	60	171	65 46	59	171	66 45	60	170
5	62 39	+60	169	63 39	+59	169	64 38	+59	169	65 37	+60	168	66 37	+59	168
6	62 31	59	167	63 30	59	167	64 29	59	167	65 28	59	166	66 27	58	166
7	62 20	59	165	63 19	59	165	64 18	58	164	65 16	59	164	66 15	58	163
8	62 08	58	163	63 06	59	163	64 05	58	162	65 03	58	162	66 01	58	161
9	61 54	58	161	62 52	58	161	63 50	58	160	64 48	58	160	65 46	57	159
10	61 39	+58	159	62 37	+57	159	63 34	+58	158	64 32	+57	157	65 29	+57	157
11	61 23	57	157	62 20	57	157	63 17	57	156	64 14	57	155	65 11	56	155
12	61 05	57	156	62 02	56	155	62 58	57	154	63 55	56	153	64 51	55	153
13	60 46	56	154	61 42	56	153	62 38	56	152	63 34	55	151	64 29	55	150
14	60 25	56	152	61 21	56	151	62 17	55	150	63 12	55	149	64 07	54	148
15	60 04	+55	150	60 59	+55	149	61 54	+54	148	62 48	+55	147	63 43	+53	147
16	59 41	54	148	60 35	55	147	61 30	54	147	62 24	53	146	63 17	54	145
17	59 17	54	146	60 11	54	146	61 05	53	145	61 58	53	144	62 51	52	143
18	58 51	54	145	59 45	53	144	60 38	53	143	61 31	52	142	62 23	52	141
19	58 25	53	143	59 18	53	142	60 11	52	141	61 03	52	140	61 55	51	139
20	57 58	+52	142	58 50	+52	141	59 42	+52	140	60 34	+51	139	61 25	+51	138
21	57 30	52	140	58 22	51	139	59 13	51	138	60 04	50	137	60 54	50	136
22	57 00	52	138	57 52	51	137	58 43	50	136	59 33	50	135	60 23	49	134
23	56 30	51	137	57 21	51	136	58 12	49	135	59 01	50	134	59 51	48	133
24	55 59	51	135	56 50	50	134	57 40	49	133	58 29	48	132	59 17	48	131
25	55 28	+49	134	56 17	+50	133	57 07	+48	132	57 55	+48	131	58 43	+48	130
26	54 55	49	133	55 44	49	132	56 33	48	131	57 21	48	129	58 09	47	128
27	54 22	49	131	55 11	48	130	55 59	47	129	56 46	47	128	57 33	47	127
28	53 48	48	130	54 36	48	129	55 24	47	128	56 11	46	127	56 57	46	126
29	53 14	47	129	54 01	47	128	54 48	47	126	55 35	46	125	56 21	45	124
30	52 38	+48	127	53 26	+46	126	54 12	+46	125	54 58	+46	124	55 44	+44	123
31	52 03	46	126	52 49	47	125	53 36	45	124	54 21	45	123	55 06	44	122
32	51 26	47	125	52 13	45	124	52 58	45	123	53 43	45	122	54 28	44	121
33	50 49	46	124	51 35	45	123	52 20	45	122	53 05	44	120	53 49	43	119
34	50 12	45	123	50 57	45	121	51 42	44	120	52 26	44	119	53 10	43	118
35	49 34	+45	121	50 19	+45	120	51 04	+43	119	51 47	+43	118	52 30	+43	117
36	48 56	44	120	49 40	44	119	50 24	44	118	51 08	43	117	51 51	42	116
37	48 17	44	119	49 01	44	118	49 45	43	117	50 28	42	116	51 10	42	115
38	47 38	44	118	48 22	43	117	49 05	43	116	49 48	42	115	50 30	41	114
39	46 58	44	117	47 42	43	116	48 25	42	115	49 07	42	114	49 49	41	113
40	46 19	+43	116	47 02	+42	115	47 44	+42	114	48 26	+41	113	49 07	+41	112
41	45 38	43	115	46 21	42	114	47 03	42	113	47 45	41	112	48 26	40	111
42	44 58	42	114	45 40	42	113	46 22	41	112	47 03	41	111	47 44	40	110
43	44 17	42	113	44 59	42	112	45 41	41	111	46 22	40	110	47 02	40	109
44	43 36	42	112	44 18	41	111	44 59	41	110	45 40	40	109	46 20	39	108
45	42 54	+42	111	43 36	+41	110	44 17	+40	109	44 57	+40	108	45 37	+39	107
46	42 13	41	110	42 54	41	109	43 35	40	108	44 15	39	107	44 54	40	106
47	41 31	41	109	42 12	40	108	42 52	40	107	43 32	40	106	44 12	38	105
48	40 48	41	109	41 29	40	108	42 09	40	107	42 49	39	106	43 28	39	105
49	40 06	41	108	40 47	40	107	41 27	39	106	42 06	39	105	42 45	39	104
50	39 23	+41	107	40 04	+40	106	40 44	+39	105	41 23	+39	104	42 02	+38	103
51	38 41	40	106	39 21	39	105	40 00	40	104	40 40	38	103	41 18	38	102
52	37 58	40	105	38 38	39	104	39 17	39	103	39 56	39	102	40 35	38	101
53	37 15	39	104	37 54	40	103	38 34	38	102	39 12	39	101	39 51	38	100
54	36 31	40	104	37 11	39	103	37 50	39	102	38 29	38	101	39 07	38	100
55	35 48	+39	103	36 27	+39	102	37 06	+39	101	37 45	+38	100	38 23	+37	99
56	35 04	40	102	35 44	38	101	36 22	39	100	37 01	38	99	37 39	37	98
57	34 21	39	101	35 00	38	100	35 38	39	99	36 17	38	98	36 55	37	97
58	33 37	39	100	34 16	38	100	34 54	39	99	35 33	37	98	36 10	38	97
59	32 53	39	100	33 32	38	99	34 10	38	98	34 48	38	97	35 26	37	96
60	32 09	+39	99	32 48	+38	98	33 26	+38	97	34 04	+38	96	34 42	+37	95
61	31 25	38	98	32 03	39	97	32 42	38	96	33 20	37	95	33 57	37	94
62	30 41	38	97	31 19	38	97	31 57	38	96	32 35	38	95	33 13	37	94
63	29 56	39	97	30 35	38	96	31 13	38	95	31 51	37	94	32 28	37	93
64	29 12	38	96	29 50	39	95	30 29	37	94	31 06	38	93	31 44	37	92
65	28 28	+38	95	29 06	+38	94	29 44	+38	94	30 22	+37	93	30 59	+37	92
66	27 43	39	95	28 22	38	94	29 00	37	93	29 37	38	92	30 15	36	91
67	26 59	38	94	27 37	38	93	28 15	38	92	28 53	37	91	29 30	37	90
68	26 14	39	93	26 53	37	92	27 30	38	91	28 08	37	91	28 45	37	90
69	25 30	38	93	26 08	38	92	26 46	37	91	27 23	38	90	28 01	37	89

LHA	20° Hc	20° d	20° Z	21° Hc	21° d	21° Z	22° Hc	22° d	22° Z	23° Hc	23° d	23° Z	24° Hc	24° d	24° Z
	° ′	′	°	° ′	′	°	° ′	′	°	° ′	′	°	° ′	′	°
0	68 00	+60	180	69 00	+60	180	70 00	+60	180	71 00	+60	180	72 00	+60	180
1	67 59	60	178	68 59	60	177	69 59	60	177	70 59	60	177	71 59	60	177
2	67 56	60	175	68 56	60	175	69 56	60	175	70 56	59	174	71 55	60	174
3	67 51	60	173	68 51	60	172	69 51	59	172	70 50	60	172	71 50	59	171
4	67 45	59	170	68 44	59	170	69 43	60	169	70 43	59	169	71 42	59	168
5	67 36	+59	168	68 35	+59	167	69 34	+59	167	70 33	+59	166	71 32	+58	165
6	67 25	59	165	68 24	59	165	69 23	58	164	70 21	58	163	71 19	59	163
7	67 13	58	163	68 11	58	162	69 09	58	162	70 07	58	161	71 05	58	160
8	66 59	58	161	67 57	57	160	68 54	58	159	69 52	57	158	70 49	57	157
9	66 43	58	158	67 41	56	157	68 37	57	157	69 34	57	156	70 31	56	155
10	66 26	+57	156	67 23	+56	155	68 19	+56	154	69 15	+56	153	70 11	+55	152
11	66 07	56	154	67 03	56	153	67 59	55	152	68 54	55	151	69 49	54	150
12	65 46	56	152	66 42	55	151	67 37	54	150	68 31	55	149	69 26	53	147
13	65 24	55	150	66 19	55	149	67 14	53	147	68 07	54	146	69 01	53	145
14	65 01	54	147	65 55	54	146	66 49	53	145	67 42	52	144	68 34	52	143
15	64 36	+54	145	65 30	+53	144	66 23	+52	143	67 15	+52	142	68 07	+51	141
16	64 11	52	144	65 03	52	142	65 55	52	141	66 47	51	140	67 38	50	139
17	63 43	52	142	64 35	52	141	65 27	51	139	66 18	50	138	67 08	49	137
18	63 15	52	140	64 07	50	139	64 57	50	137	65 47	50	136	66 37	48	135
19	62 46	51	138	63 37	50	137	64 27	49	136	65 16	48	134	66 04	48	133
20	62 16	+50	136	63 06	+49	135	63 55	+49	134	64 44	+47	133	65 31	+47	131
21	61 44	50	135	62 34	48	133	63 22	48	132	64 10	47	131	64 57	46	129
22	61 12	49	133	62 01	48	132	62 49	47	131	63 36	46	129	64 22	46	128
23	60 39	48	132	61 27	48	130	62 15	46	129	63 01	46	128	63 47	44	126
24	60 05	48	130	60 53	46	129	61 39	46	127	62 25	45	126	63 10	44	125
25	59 31	+47	129	60 18	+46	127	61 04	+45	126	61 49	+44	125	62 33	+44	123
26	58 56	46	127	59 42	45	126	60 27	45	125	61 12	44	123	61 56	42	122
27	58 20	45	126	59 05	45	125	59 50	44	123	60 34	43	122	61 17	43	120
28	57 43	45	124	58 28	45	123	59 13	43	122	59 56	43	120	60 39	41	119
29	57 06	45	123	57 51	43	122	58 34	43	120	59 17	42	119	59 59	42	118
30	56 28	+44	122	57 12	+44	121	57 56	+42	119	58 38	+42	118	59 20	+40	116
31	55 50	44	121	56 34	43	119	57 17	42	118	57 59	41	117	58 40	40	115
32	55 12	43	119	55 55	42	118	56 37	41	117	57 18	41	115	57 59	40	114
33	54 32	43	118	55 15	42	117	55 57	41	116	56 38	40	114	57 18	39	113
34	53 53	42	117	54 35	42	116	55 17	40	115	55 57	40	113	56 37	39	112
35	53 13	+42	116	53 55	+41	115	54 36	+40	113	55 16	+39	112	55 55	+39	111
36	52 33	41	115	53 14	41	114	53 55	40	112	54 35	39	111	55 14	38	110
37	51 52	41	114	52 33	40	113	53 13	40	111	53 53	38	110	54 31	38	109
38	51 11	41	113	51 52	40	112	52 32	39	110	53 11	38	109	53 49	38	108
39	50 30	40	112	51 10	40	111	51 50	38	109	52 28	38	108	53 06	38	107
40	49 48	+40	111	50 28	+39	110	51 07	+39	108	51 46	+38	107	52 24	+37	106
41	49 06	40	110	49 46	39	109	50 25	38	107	51 03	38	106	51 41	36	105
42	48 24	40	109	49 04	38	108	49 42	38	106	50 20	37	105	50 57	37	104
43	47 42	39	108	48 21	38	107	48 59	38	106	49 37	37	104	50 14	36	103
44	46 59	39	107	47 38	38	106	48 16	38	105	48 54	37	103	49 31	36	102
45	46 16	+39	106	46 55	+38	105	47 33	+37	104	48 10	+37	103	48 47	+36	101
46	45 34	38	105	46 12	38	104	46 50	37	103	47 27	36	102	48 03	36	101
47	44 50	39	104	45 29	37	103	46 06	37	102	46 43	36	101	47 19	36	100
48	44 07	38	103	44 45	38	102	45 23	36	101	45 59	36	100	46 35	36	99
49	43 24	37	103	44 01	38	102	44 39	36	100	45 15	36	99	45 51	35	98
50	42 40	+38	102	43 18	+37	101	43 55	+36	100	44 31	+36	99	45 07	+35	97
51	41 56	38	101	42 34	37	100	43 11	36	99	43 47	36	98	44 23	35	97
52	41 13	37	100	41 50	37	99	42 27	36	98	43 03	35	97	43 38	35	96
53	40 29	37	99	41 06	37	98	41 43	36	97	42 19	35	96	42 54	35	95
54	39 45	37	99	40 22	36	98	40 58	36	97	41 34	36	96	42 10	34	94
55	39 00	+37	98	39 37	+37	97	40 14	+36	96	40 50	+35	95	41 25	+35	94
56	38 16	37	97	38 53	37	96	39 30	35	95	40 05	36	94	40 41	34	93
57	37 32	37	96	38 09	36	95	38 45	36	94	39 21	35	93	39 56	35	93
58	36 48	36	96	37 24	37	95	38 01	35	94	38 36	36	93	39 12	34	92
59	36 03	37	95	36 40	36	94	37 16	36	93	37 52	35	92	38 27	35	91
60	35 19	+36	94	35 55	+37	93	36 32	+35	92	37 07	+35	91	37 42	+35	90
61	34 34	37	93	35 11	36	93	35 47	36	92	36 23	35	91	36 58	34	90
62	33 50	36	93	34 26	36	92	35 02	36	91	35 38	35	90	36 13	35	89
63	33 05	37	92	33 42	36	91	34 18	35	90	34 53	36	89	35 29	34	88
64	32 21	36	91	32 57	36	91	33 33	36	90	34 09	35	89	34 44	35	88
65	31 36	+37	91	32 13	+36	90	32 49	+35	89	33 24	+36	88	34 00	+34	87
66	30 51	37	90	31 28	36	89	32 04	36	88	32 40	35	87	33 15	35	87
67	30 07	36	89	30 43	37	89	31 20	35	88	31 55	36	87	32 31	34	86
68	29 22	37	89	29 59	36	88	30 35	36	87	31 11	35	86	31 46	35	85
69	28 38	36	88	29 14	36	87	29 50	36	86	30 26	36	85	31 02	35	84

LHA	25° Hc	25° d	25° Z	26° Hc	26° d	26° Z	27° Hc	27° d	27° Z	28° Hc	28° d	28° Z	29° Hc	29° d	29° Z	LHA
	° ′	′	°	° ′	′	°	° ′	′	°	° ′	′	°	° ′	′	°	
0	73 00	+60	180	74 00	+60	180	75 00	+60	180	76 00	+60	180	77 00	+60	180	360
1	72 59	60	177	73 59	60	177	74 59	60	177	75 59	60	176	76 59	59	176	359
2	72 55	60	174	73 55	60	174	74 55	59	173	75 54	60	173	76 54	60	172	358
3	72 49	60	171	73 49	59	170	74 48	59	170	75 47	60	169	76 47	59	169	357
4	72 41	59	168	73 40	59	167	74 39	58	166	75 38	58	166	76 36	59	165	356
5	72 30	+59	165	73 29	+58	164	74 27	+58	163	75 25	+58	162	76 23	+58	161	355
6	72 18	57	162	73 15	58	161	74 13	57	160	75 10	57	159	76 07	57	158	354
7	72 03	57	159	73 00	57	158	73 57	56	157	74 53	56	156	75 49	56	154	353
8	71 46	56	156	72 42	56	155	73 38	56	154	74 34	55	153	75 29	54	151	352
9	71 27	55	154	72 22	55	152	73 17	55	151	74 12	54	150	75 06	53	148	351
10	71 06	+55	151	72 01	+54	150	72 55	+53	148	73 48	+53	147	74 41	+52	145	350
11	70 43	54	148	71 37	54	147	72 31	52	146	73 23	52	144	74 15	50	142	349
12	70 19	53	146	71 12	53	145	72 05	51	143	72 56	50	141	73 46	50	139	348
13	69 54	52	144	70 46	51	142	71 37	50	141	72 27	50	139	73 17	48	137	347
14	69 26	52	141	70 18	50	140	71 08	49	138	71 57	49	136	72 46	47	134	346
15	68 58	+50	139	69 48	+50	138	70 38	+48	136	71 26	+47	134	72 13	+46	132	345
16	68 28	49	137	69 17	49	136	70 06	47	134	70 53	46	132	71 39	45	130	344
17	67 57	49	135	68 46	47	134	69 33	47	132	70 20	45	130	71 05	44	128	343
18	67 25	48	133	68 13	46	132	68 59	46	130	69 45	44	128	70 29	43	126	342
19	66 52	47	131	67 39	46	130	68 25	44	128	69 09	44	126	69 53	42	124	341
20	66 18	+46	130	67 04	+45	128	67 49	+44	126	68 33	+42	124	69 15	+41	122	340
21	65 43	46	128	66 29	44	126	67 13	43	125	67 56	41	123	68 37	41	121	339
22	65 08	44	126	65 52	44	125	66 36	42	123	67 18	41	121	67 59	39	119	338
23	64 31	44	125	65 15	43	123	65 58	41	121	66 39	40	120	67 19	39	118	337
24	63 54	43	123	64 37	42	122	65 19	41	120	66 00	40	118	66 40	38	116	336
25	63 17	+42	122	63 59	+41	120	64 40	+41	118	65 21	+38	117	65 59	+38	115	335
26	62 38	42	120	63 20	41	119	64 01	39	117	64 40	39	115	65 19	37	113	334
27	62 00	41	119	62 41	40	117	63 21	39	116	64 00	38	114	64 38	36	112	333
28	61 20	41	118	62 01	40	116	62 41	38	114	63 19	37	113	63 56	36	111	332
29	60 41	40	116	61 21	39	115	62 00	38	113	62 38	36	111	63 14	36	110	331
30	60 00	+40	115	60 40	+39	114	61 19	+37	112	61 56	+36	110	62 32	+35	109	330
31	59 20	39	114	59 59	38	112	60 37	37	111	61 14	36	109	61 50	34	107	329
32	58 39	38	113	59 17	38	111	59 55	37	110	60 32	35	108	61 07	34	106	328
33	57 57	39	112	58 36	37	110	59 13	36	109	59 49	35	107	60 24	34	105	327
34	57 16	38	110	57 54	37	109	58 31	35	108	59 06	35	106	59 41	33	104	326
35	56 34	+37	109	57 11	+37	108	57 48	+35	107	58 23	+35	105	58 58	+33	103	325
36	55 52	37	108	56 29	36	107	57 05	35	106	57 40	34	104	58 14	33	102	324
37	55 09	37	107	55 46	36	106	56 22	35	105	56 57	34	103	57 31	32	102	323
38	54 27	36	106	55 03	36	105	55 39	34	104	56 13	34	102	56 47	32	101	322
39	53 44	36	105	54 20	35	104	54 55	35	103	55 30	33	101	56 03	32	100	321
40	53 01	+36	105	53 37	+35	103	54 12	+34	102	54 46	+33	100	55 19	+32	99	320
41	52 17	36	104	52 53	35	102	53 28	34	101	54 02	33	100	54 35	32	98	319
42	51 34	35	103	52 09	35	101	52 44	34	100	53 18	33	99	53 51	32	97	318
43	50 50	36	102	51 26	34	101	52 00	34	99	52 34	32	98	53 06	32	97	317
44	50 07	35	101	50 42	34	100	51 16	34	98	51 50	32	97	52 22	32	96	316
45	49 23	+35	100	49 58	+34	99	50 32	+33	98	51 05	+33	96	51 38	+31	95	315
46	48 39	35	99	49 14	34	98	49 48	33	97	50 21	32	96	50 53	32	94	314
47	47 55	35	99	48 30	33	97	49 03	33	96	49 37	32	95	50 09	31	94	313
48	47 11	34	98	47 45	34	97	48 19	33	95	48 52	32	94	49 24	32	93	312
49	46 26	35	97	47 01	34	96	47 35	33	95	48 08	32	93	48 40	31	92	311
50	45 42	+35	96	46 17	+33	95	46 50	+33	94	47 23	+32	92	47 55	+31	91	310
51	44 58	34	96	45 32	34	94	46 06	33	93	46 39	32	92	47 11	31	91	309
52	44 13	35	95	44 48	33	94	45 21	33	92	45 54	32	91	46 26	31	90	308
53	43 29	34	94	44 03	34	93	44 37	32	92	45 09	32	91	45 41	32	89	307
54	42 44	35	93	43 19	33	92	43 52	33	91	44 25	32	90	44 57	31	89	306
55	42 00	+34	93	42 34	+33	91	43 07	+33	90	43 40	+32	89	44 12	+32	88	305
56	41 15	34	92	41 49	34	91	42 23	33	90	42 56	32	89	43 28	31	87	304
57	40 31	34	91	41 05	33	90	41 38	33	89	42 11	32	88	42 43	32	87	303
58	39 46	34	91	40 20	33	90	40 54	33	88	41 27	32	87	41 59	31	86	302
59	39 02	34	90	39 36	33	89	40 09	33	88	40 42	32	87	41 14	32	86	301
60	38 17	+34	89	38 51	+34	88	39 25	+32	87	39 57	+33	86	40 30	+31	85	300
61	37 32	35	89	38 07	33	88	38 40	33	87	39 13	32	85	39 45	32	84	299
62	36 48	34	88	37 22	34	87	37 56	33	86	38 29	32	85	39 01	32	84	298
63	36 03	34	87	36 37	34	86	37 11	33	85	37 44	33	84	38 17	32	83	297
64	35 19	34	87	35 53	34	86	36 27	33	85	37 00	33	84	37 33	32	83	296
65	34 34	+35	86	35 09	+33	85	35 42	+34	84	36 16	+32	83	36 48	+32	82	295
66	33 50	34	85	34 24	34	84	34 58	33	83	35 31	33	82	36 04	33	81	294
67	33 05	35	85	33 40	34	84	34 14	33	83	34 47	33	82	35 20	33	81	293
68	32 21	35	84	32 56	34	83	33 30	33	82	34 03	33	81	34 36	33	80	292
69	31 37	34	84	32 11	34	83	32 45	34	82	33 19	33	81	33 52	33	80	291

S. Lat. {LHA greater than 180°....... Zn=180−Z / LHA less than 180°.......... Zn=180+Z}

DECLINATION (15°-29°) SAME NAME AS LATITUDE

LAT 42°

N. Lat. {LHA greater than 180°........ Zn=Z; LHA less than 180°........... Zn=360−Z}

DECLINATION (0°-14°) CONTRARY NAME TO LATITUDE

LHA	0° Hc	d	Z	1° Hc	d	Z	2° Hc	d	Z	3° Hc	d	Z	4° Hc	d	Z
	° ′	′	°	° ′	′	°	° ′	′	°	° ′	′	°	° ′	′	°
69	15 27	42	104	14 45	42	105	14 03	42	106	13 21	42	107	12 39	42	107
68	16 10	42	105	15 28	42	106	14 46	42	107	14 04	42	107	13 22	43	108
67	16 53	42	106	16 11	42	107	15 29	43	107	14 46	42	108	14 04	43	109
66	17 36	43	107	16 53	42	107	16 11	42	108	15 29	43	109	14 46	43	110
65	18 18	−42	107	17 36	−43	108	16 53	−42	109	16 11	−43	110	15 28	−43	110
64	19 01	43	108	18 18	42	109	17 36	43	110	16 53	43	110	16 10	44	111
63	19 43	43	109	19 00	43	110	18 17	43	110	17 34	43	111	16 51	43	112
62	20 25	43	110	19 42	43	110	18 59	43	111	18 16	44	112	17 32	43	113
61	21 07	43	110	20 24	43	111	19 41	44	112	18 57	44	113	18 13	43	113
60	21 49	−44	111	21 05	−43	112	20 22	−44	113	19 38	−44	113	18 54	−44	114
59	22 30	43	112	21 47	44	113	21 03	44	113	20 19	44	114	19 35	44	115
58	23 12	44	113	22 28	44	113	21 44	44	114	21 00	45	115	20 15	44	116
57	23 53	44	114	23 09	45	114	22 24	44	115	21 40	45	116	20 55	44	116
56	24 33	44	114	23 49	44	115	23 05	45	116	22 20	45	117	21 35	45	117
55	25 14	−45	115	24 29	−44	116	23 45	−45	117	23 00	−45	117	22 15	−46	118
54	25 54	45	116	25 09	45	117	24 24	45	117	23 39	45	118	22 54	46	119
53	26 34	45	117	25 49	45	118	25 04	46	118	24 18	45	119	23 33	46	120
52	27 14	46	118	26 28	45	118	25 43	46	119	24 57	46	120	24 11	46	121
51	27 53	45	119	27 08	46	119	26 22	46	120	25 36	46	121	24 50	47	121
50	28 32	−46	119	27 46	−46	120	27 00	−46	121	26 14	−46	122	25 28	−47	122
49	29 11	46	120	28 25	47	121	27 38	46	122	26 52	47	122	26 05	47	123
48	29 49	46	121	29 03	47	122	28 16	47	123	27 29	47	123	26 42	47	124
47	30 27	46	122	29 41	47	123	28 54	48	123	28 06	47	124	27 19	47	125
46	31 05	47	123	30 18	47	124	29 31	48	124	28 43	48	125	27 55	47	126
45	31 42	−47	124	30 55	−48	125	30 07	−48	125	29 19	−48	126	28 31	−48	127
44	32 19	48	125	31 31	48	125	30 43	48	126	29 55	48	127	29 07	48	128
43	32 55	48	126	32 07	48	126	31 19	48	127	30 31	49	128	29 42	49	128
42	33 31	48	127	32 43	48	127	31 55	49	128	31 06	49	129	30 17	49	129
41	34 07	49	128	33 18	48	128	32 30	50	129	31 40	49	130	30 51	49	130
40	34 42	−49	129	33 53	−49	129	33 04	−49	130	32 15	−50	131	31 25	−50	131
39	35 17	50	130	34 27	49	130	33 38	50	131	32 48	50	132	31 58	50	132
38	35 51	50	131	35 01	50	131	34 11	50	132	33 21	50	133	32 31	51	133
37	36 24	50	132	35 34	50	132	34 44	50	133	33 54	51	134	33 03	51	134
36	36 57	50	133	36 07	50	133	35 17	51	134	34 26	51	135	33 35	51	135
35	37 30	−51	134	36 39	−51	134	35 48	−51	135	34 57	−51	136	34 06	−52	136
34	38 02	51	135	37 11	51	135	36 20	52	136	35 28	52	137	34 36	52	137
33	38 33	51	136	37 42	52	137	36 50	52	137	35 58	52	138	35 06	52	138
32	39 04	52	137	38 12	52	138	37 20	52	138	36 28	52	139	35 36	53	140
31	39 34	52	138	38 42	52	139	37 50	53	139	36 57	53	140	36 04	53	141
30	40 04	−53	139	39 11	−53	140	38 18	−53	140	37 25	−53	141	36 32	−53	142
29	40 32	53	140	39 39	53	141	38 46	53	142	37 53	53	142	37 00	54	143
28	41 00	53	142	40 07	53	142	39 14	54	143	38 20	54	143	37 26	54	144
27	41 28	54	143	40 34	54	143	39 40	54	144	38 46	54	144	37 52	54	145
26	41 55	55	144	41 00	54	145	40 06	54	145	39 12	55	146	38 17	54	146
25	42 20	−54	145	41 26	−55	146	40 31	−54	146	39 37	−55	147	38 42	−55	147
24	42 45	54	146	41 51	55	147	40 56	55	148	40 01	55	148	39 06	56	149
23	43 10	55	148	42 15	56	148	41 19	55	149	40 24	56	149	39 28	55	150
22	43 33	55	149	42 38	56	149	41 42	56	150	40 46	55	150	39 51	56	151
21	43 56	56	150	43 00	56	151	42 04	56	151	41 08	56	152	40 12	56	152
20	44 18	−57	152	43 21	−56	152	42 25	−56	152	41 29	−57	153	40 32	−56	153
19	44 38	56	153	43 42	57	153	42 45	56	154	41 49	57	154	40 52	57	155
18	44 58	56	154	44 02	57	155	43 05	57	155	42 08	57	155	41 11	58	156
17	45 17	57	155	44 20	57	156	43 23	57	156	42 26	58	157	41 28	57	157
16	45 35	57	157	44 38	57	157	43 41	58	158	42 43	58	158	41 45	57	158
15	45 53	−58	158	44 55	−58	159	43 57	−58	159	42 59	−58	159	42 01	−58	160
14	46 09	58	160	45 11	58	160	44 13	59	160	43 14	58	161	42 16	58	161
13	46 24	59	161	45 25	58	161	44 27	58	162	43 29	59	162	42 30	58	162
12	46 38	59	162	45 39	58	163	44 41	59	163	43 42	59	163	42 43	58	164
11	46 51	59	164	45 52	59	164	44 53	59	164	43 54	58	165	42 56	59	165
10	47 03	−59	165	46 04	−59	166	45 05	−59	166	44 06	−59	166	43 07	−59	166
9	47 13	59	167	46 14	59	167	45 15	59	167	44 16	59	167	43 17	60	168
8	47 23	59	168	46 24	60	168	45 24	59	169	44 25	59	169	43 26	60	169
7	47 32	60	170	46 32	59	170	45 33	60	170	44 33	59	170	43 34	60	170
6	47 39	59	171	46 40	60	171	45 40	60	171	44 40	59	172	43 41	60	172
5	47 46	−60	173	46 46	−60	173	45 46	−60	173	44 46	−59	173	43 47	−60	173
4	47 51	60	174	46 51	60	174	45 51	60	174	44 51	60	174	43 51	59	175
3	47 55	60	176	46 55	60	176	45 55	60	176	44 55	60	176	43 55	60	176
2	47 58	60	177	46 58	60	177	45 58	60	177	44 58	60	177	43 58	60	177
1	47 59	60	179	46 59	60	179	45 59	59	179	45 00	60	179	44 00	60	179
0	48 00	−60	180	47 00	−60	180	46 00	−60	180	45 00	−60	180	44 00	−60	180

LHA	5° Hc	d	Z	6° Hc	d	Z	7° Hc	d	Z	8° Hc	d	Z	9° Hc	d	Z
	° ′	′	°	° ′	′	°	° ′	′	°	° ′	′	°	° ′	′	°
69	11 57	43	108	11 14	42	109	10 32	43	110	09 49	42	110	09 07	43	111
68	12 39	42	109	11 57	43	110	11 14	43	110	10 31	43	111	09 48	43	112
67	13 21	43	110	12 38	42	110	11 56	43	111	11 13	43	112	10 30	44	112
66	14 03	43	110	13 20	43	111	12 37	43	112	11 54	43	112	11 11	44	113
65	14 45	−43	111	14 02	−43	112	13 19	−44	112	12 35	−43	113	11 52	−44	114
64	15 26	43	112	14 43	43	113	14 00	44	113	13 16	44	114	12 32	43	115
63	16 08	44	113	15 24	44	113	14 40	43	114	13 57	44	115	13 13	44	115
62	16 49	44	113	16 05	44	114	15 21	44	115	14 37	44	115	13 53	44	116
61	17 30	44	114	16 46	44	115	16 02	45	115	15 17	44	116	14 33	45	117
60	18 10	−44	115	17 26	−44	116	16 42	−45	116	15 57	−44	117	15 13	−45	118
59	18 51	45	116	18 06	44	116	17 22	45	117	16 37	45	118	15 52	45	118
58	19 31	45	116	18 46	45	117	18 01	45	118	17 16	45	118	16 31	45	119
57	20 11	45	117	19 26	46	118	18 40	45	119	17 55	45	119	17 10	46	120
56	20 50	45	118	20 05	45	119	19 20	46	119	18 34	46	120	17 48	45	121
55	21 29	−45	119	20 44	−46	119	19 58	−46	120	19 12	−45	121	18 27	−47	122
54	22 08	45	120	21 23	46	120	20 37	46	121	19 51	47	122	19 04	46	122
53	22 47	46	120	22 01	46	121	21 15	47	122	20 28	46	122	19 42	47	123
52	23 25	46	121	22 39	47	122	21 52	46	123	21 06	47	123	20 19	47	124
51	24 03	46	122	23 17	47	123	22 30	47	123	21 43	47	124	20 56	47	125
50	24 41	−47	123	23 54	−47	124	23 07	−47	124	22 20	−48	125	21 32	−47	126
49	25 18	47	124	24 31	47	124	23 44	48	125	22 56	48	126	22 08	47	126
48	25 55	47	125	25 08	48	125	24 20	48	126	23 32	48	127	22 44	48	127
47	26 32	48	126	25 44	48	126	24 56	48	127	24 08	49	128	23 19	48	128
46	27 08	48	126	26 20	49	127	25 31	48	128	24 43	49	128	23 54	48	129
45	27 43	−48	127	26 55	−49	128	26 06	−48	129	25 18	−49	129	24 29	−49	130
44	28 19	49	128	27 30	49	129	26 41	49	130	25 52	49	130	25 03	50	131
43	28 53	49	129	28 04	49	130	27 15	49	130	26 26	50	131	25 36	49	132
42	29 28	50	130	28 38	49	131	27 49	50	131	26 59	50	132	26 09	50	133
41	30 02	50	131	29 12	50	132	28 22	50	132	27 32	50	133	26 42	50	134
40	30 35	−50	132	29 45	−50	133	28 55	−50	133	28 05	−51	134	27 14	−51	134
39	31 08	50	133	30 18	51	134	29 27	51	134	28 36	50	135	27 46	51	135
38	31 40	50	134	30 50	51	135	29 59	51	135	29 08	51	136	28 17	52	136
37	32 12	51	135	31 21	51	136	30 30	51	136	29 39	52	137	28 47	52	137
36	32 44	52	136	31 52	51	137	31 01	52	137	30 09	52	138	29 17	52	138
35	33 14	−51	137	32 23	−52	138	31 31	−52	138	30 39	−53	139	29 46	−52	139
34	33 44	52	138	32 52	52	139	32 00	52	139	31 08	53	140	30 15	52	140
33	34 14	52	139	33 22	53	140	32 29	53	140	31 36	53	141	30 43	53	141
32	34 43	53	140	33 50	53	141	32 57	53	141	32 04	53	142	31 11	53	142
31	35 11	53	141	34 18	53	142	33 25	53	142	32 32	54	143	31 38	54	143
30	35 39	−53	142	34 46	−54	143	33 52	−54	143	32 58	−54	144	32 04	−54	144
29	36 06	54	143	35 12	54	144	34 18	54	144	33 24	54	145	32 30	54	145
28	36 32	54	144	35 38	54	145	34 44	54	146	33 50	55	146	32 55	55	147
27	36 58	55	146	36 03	54	146	35 09	55	147	34 14	55	147	33 19	55	148
26	37 23	55	147	36 28	55	147	35 33	55	148	34 38	55	148	33 43	55	149
25	37 47	−55	148	36 52	−55	148	35 57	−56	149	35 01	−55	149	34 06	−56	150
24	38 10	55	149	37 15	56	150	36 19	55	150	35 24	56	150	34 28	56	151
23	38 33	56	150	37 37	56	151	36 41	56	151	35 45	56	152	34 49	56	152
22	38 55	56	151	37 59	57	152	37 02	56	152	36 06	56	153	35 10	57	153
21	39 16	57	153	38 19	56	153	37 23	57	153	36 26	57	154	35 29	56	154
20	39 36	−57	154	38 39	−57	154	37 42	−57	155	36 45	−57	155	35 48	−57	155
19	39 55	57	155	38 58	57	155	38 01	57	156	37 04	57	156	36 07	58	157
18	40 13	57	156	39 16	57	157	38 19	58	157	37 21	57	157	36 24	58	158
17	40 31	58	158	39 33	57	158	38 36	58	158	37 38	58	159	36 40	57	159
16	40 48	58	159	39 50	58	159	38 52	58	159	37 54	58	160	36 56	58	160
15	41 03	−58	160	40 05	−58	160	39 07	−58	161	38 09	−58	161	37 11	−58	161
14	41 18	58	161	40 20	59	162	39 21	58	162	38 23	58	162	37 25	59	163
13	41 32	59	163	40 33	58	163	39 35	59	163	38 36	58	163	37 38	59	164
12	41 45	59	164	40 46	59	164	39 47	58	164	38 49	59	165	37 50	59	165
11	41 57	59	165	40 58	59	165	39 59	59	166	39 00	59	166	38 01	59	166
10	42 08	−60	167	41 08	−59	167	40 09	−59	167	39 10	−59	167	38 11	−59	167
9	42 17	59	168	41 18	59	168	40 19	59	168	39 20	60	168	38 20	59	169
8	42 26	59	169	41 27	60	169	40 27	59	170	39 28	59	170	38 29	60	170
7	42 34	59	171	41 35	60	171	40 35	59	171	39 36	60	171	38 36	60	171
6	42 41	60	172	41 41	59	172	40 42	60	172	39 42	60	172	38 42	59	172
5	42 47	−60	173	41 47	−60	173	40 47	−59	173	39 48	−60	174	38 48	−60	174
4	42 52	60	175	41 52	60	175	40 52	60	175	39 52	60	175	38 52	60	175
3	42 55	60	176	41 55	60	176	40 55	59	176	39 56	60	176	38 56	60	176
2	42 58	60	177	41 58	60	177	40 58	60	177	39 58	60	177	38 58	60	178
1	43 00	60	179	42 00	60	179	41 00	60	179	40 00	60	179	39 00	60	179
0	43 00	−60	180	42 00	−60	180	41 00	−60	180	40 00	−60	180	39 00	−60	180

LHA	10° Hc	d	Z	11° Hc	d	Z	12° Hc	d	Z	13° Hc	d	Z	14° Hc	d	Z	LHA
	° ′	′	°	° ′	′	°	° ′	′	°	° ′	′	°	° ′	′	°	
69	08 24	43	112	07 41	43	112	06 58	43	113	06 15	43	114	05 32	43	115	291
68	09 05	43	112	08 22	43	113	07 39	43	114	06 56	43	115	06 13	43	115	292
67	09 46	43	113	09 03	43	114	08 20	43	115	07 37	44	115	06 53	43	116	293
66	10 27	43	114	09 44	44	115	09 00	43	115	08 17	44	116	07 33	44	117	294
65	11 08	−44	115	10 24	−43	115	09 41	−44	116	08 57	−44	117	08 13	−44	117	295
64	11 49	44	115	11 05	44	116	10 21	44	117	09 37	45	117	08 52	44	118	296
63	12 29	44	116	11 45	45	117	11 00	44	117	10 16	44	118	09 32	45	119	297
62	13 09	45	117	12 24	44	117	11 40	45	118	10 55	44	119	10 11	45	120	298
61	13 48	44	118	13 04	45	118	12 19	45	119	11 34	45	120	10 49	45	120	299
60	14 28	−45	118	13 43	−45	119	12 58	−45	120	12 13	−45	120	11 28	−46	121	300
59	15 07	45	119	14 22	46	120	13 36	45	120	12 51	45	121	12 06	46	122	301
58	15 46	46	120	15 00	45	121	14 15	46	121	13 29	46	122	12 43	45	123	302
57	16 24	45	121	15 39	46	121	14 53	46	122	14 07	46	123	13 21	46	123	303
56	17 03	46	121	16 17	47	122	15 30	46	123	14 44	46	123	13 58	46	124	304
55	17 40	−46	122	16 54	−46	123	16 08	−47	124	15 21	−46	124	14 35	−47	125	305
54	18 18	46	123	17 32	47	124	16 45	47	124	15 58	47	125	15 11	47	126	306
53	18 55	47	124	18 08	46	124	17 22	48	125	16 34	47	126	15 47	47	126	307
52	19 32	47	125	18 45	47	125	17 58	47	126	17 11	48	127	16 23	48	127	308
51	20 09	48	125	19 21	47	126	18 34	48	127	17 46	48	127	16 58	48	128	309
50	20 45	−48	126	19 57	−48	127	19 09	−48	128	18 21	−48	128	17 33	−48	129	310
49	21 21	48	127	20 33	48	128	19 45	49	128	18 56	48	129	18 08	49	130	311
48	21 56	48	128	21 08	49	129	20 19	48	129	19 31	49	130	18 42	49	130	312
47	22 31	49	129	21 42	48	129	20 54	49	130	20 05	49	131	19 16	49	131	313
46	23 06	49	130	22 17	49	130	21 28	50	131	20 38	49	132	19 49	49	132	314
45	23 40	−50	131	22 50	−49	131	22 01	−49	132	21 12	−50	132	20 22	−50	133	315
44	24 13	49	131	23 24	50	132	22 34	50	133	21 44	50	133	20 54	50	134	316
43	24 47	50	132	23 57	50	133	23 07	50	134	22 17	51	134	21 26	50	135	317
42	25 19	50	133	24 29	50	134	23 39	51	134	22 48	50	135	21 58	51	136	318
41	25 52	51	134	25 01	51	135	24 10	50	135	23 20	51	136	22 29	51	137	319
40	26 23	−51	135	25 32	−51	136	24 41	−51	136	23 50	−51	137	22 59	−51	137	320
39	26 55	52	136	26 03	51	137	25 12	51	137	24 21	52	138	23 29	52	138	321
38	27 25	51	137	26 34	52	138	25 42	52	138	24 50	51	139	23 59	52	139	322
37	27 55	51	138	27 04	52	138	26 12	52	139	25 20	53	140	24 27	52	140	323
36	28 25	52	139	27 33	52	139	26 41	53	140	25 48	52	141	24 56	53	141	324
35	28 54	−52	140	28 02	−53	140	27 09	−53	141	26 16	−53	141	25 23	−52	142	325
34	29 23	53	141	28 30	53	141	27 37	53	142	26 44	53	142	25 51	54	143	326
33	29 50	53	142	28 57	53	142	28 04	53	143	27 11	54	143	26 17	53	144	327
32	30 18	54	143	29 24	53	143	28 31	54	144	27 37	54	144	26 43	54	145	328
31	30 44	53	144	29 51	54	144	28 57	54	145	28 03	54	145	27 09	55	146	329
30	31 10	−54	145	30 16	−54	145	29 22	−54	146	28 28	−55	146	27 33	−54	147	330
29	31 36	55	146	30 41	54	146	29 47	55	147	28 52	55	147	27 57	54	148	331
28	32 00	54	147	31 06	55	147	30 11	55	148	29 16	55	148	28 21	55	149	332
27	32 24	55	148	31 29	55	149	30 34	55	149	29 39	55	149	28 44	56	150	333
26	32 48	56	149	31 52	55	150	30 57	56	150	30 01	55	150	29 06	56	151	334
25	33 10	−56	150	32 14	−55	151	31 19	−56	151	30 23	−56	152	29 27	−56	152	335
24	33 32	56	151	32 36	56	152	31 40	56	152	30 44	56	153	29 48	57	153	336
23	33 53	56	152	32 57	57	153	32 00	56	153	31 04	57	154	30 07	56	154	337
22	34 13	56	154	33 17	57	154	32 20	57	154	31 23	56	155	30 27	57	155	338
21	34 33	57	155	33 36	57	155	32 39	57	155	31 42	57	156	30 45	57	156	339
20	34 51	−57	156	33 54	−57	156	32 57	−57	157	32 00	−57	157	31 03	−58	157	340
19	35 09	57	157	34 12	57	157	33 15	58	158	32 17	57	158	31 20	58	158	341
18	35 26	57	158	34 29	58	158	33 31	58	159	32 33	57	159	31 36	58	159	342
17	35 43	58	159	34 45	58	160	33 47	58	160	32 49	58	160	31 51	58	161	343
16	35 58	58	160	35 00	58	161	34 02	58	161	33 04	59	161	32 05	58	162	344
15	36 13	−59	162	35 14	−58	162	34 16	−58	162	33 18	−59	162	32 19	−58	163	345
14	36 26	58	163	35 28	59	163	34 29	58	163	33 31	59	164	32 32	59	164	346
13	36 39	59	164	35 40	58	164	34 42	59	165	33 43	59	165	32 44	59	165	347
12	36 51	59	165	35 52	59	165	34 53	59	166	33 54	59	166	32 55	59	166	348
11	37 02	59	166	36 03	59	167	35 04	59	167	34 05	60	167	33 05	59	167	349
10	37 12	−59	168	36 13	−60	168	35 13	−59	168	34 14	−59	168	33 15	−59	168	350
9	37 21	59	169	36 22	60	169	35 22	59	169	34 23	60	169	33 23	59	170	351
8	37 29	59	170	36 30	60	170	35 30	59	170	34 31	60	171	33 31	60	171	352
7	37 36	59	171	36 37	60	171	35 37	60	172	34 37	59	172	33 38	60	172	353
6	37 43	60	173	36 43	60	173	35 43	60	173	34 43	59	173	33 44	60	173	354
5	37 48	−60	174	36 48	−60	174	35 48	−60	174	34 48	−59	174	33 49	−60	174	355
4	37 52	60	175	36 52	59	175	35 53	60	175	34 53	60	175	33 53	60	175	356
3	37 56	60	176	36 56	60	176	35 56	60	176	34 56	60	176	33 56	60	177	357
2	37 58	60	178	36 58	60	178	35 58	60	178	34 58	60	178	33 58	60	178	358
1	38 00	60	179	37 00	60	179	36 00	60	179	35 00	60	179	34 00	60	179	359
0	38 00	−60	180	37 00	−60	180	36 00	−60	180	35 00	−60	180	34 00	−60	180	360

S. Lat. {LHA greater than 180°........ Zn=180−Z; LHA less than 180°........... Zn=180+Z}

DECLINATION (0°-14°) CONTRARY NAME TO LATITUDE

N. Lat. {LHA greater than 180°....... Zn=Z
{LHA less than 180°.......... Zn=360−Z

DECLINATION (15°-29°) SAME NAME AS LATITUDE

LHA	15° Hc	15° d	15° Z	16° Hc	16° d	16° Z	17° Hc	17° d	17° Z	18° Hc	18° d	18° Z	19° Hc	19° d	19° Z
	° ′	′	°	° ′	′	°	° ′	′	°	° ′	′	°	° ′	′	°
0	62 00	+60	180	63 00	+60	180	64 00	+60	180	65 00	+60	180	66 00	+60	180
1	61 59	60	178	62 59	60	178	63 59	60	178	64 59	60	178	65 59	60	178
2	61 57	60	176	62 57	60	176	63 57	60	176	64 57	59	176	65 56	60	175
3	61 53	60	174	62 53	60	174	63 53	59	174	64 52	60	173	65 52	60	173
4	61 47	60	172	62 47	60	172	63 47	59	171	64 46	60	171	65 46	59	171
5	61 40	+60	170	62 40	+59	170	63 39	+60	169	64 39	+59	169	65 38	+59	169
6	61 32	59	168	62 31	59	167	63 30	59	167	64 29	59	167	65 28	59	166
7	61 22	59	166	62 21	59	165	63 20	58	165	64 18	59	165	65 17	59	164
8	61 10	59	164	62 09	58	163	63 07	59	163	64 06	58	162	65 04	58	162
9	60 57	59	162	61 56	58	161	62 54	58	161	63 52	58	160	64 50	57	160
10	60 43	+58	160	61 41	+58	159	62 39	+57	159	63 36	+58	158	64 34	+57	158
11	60 27	58	158	61 25	57	158	62 22	57	157	63 19	57	156	64 16	57	155
12	60 10	57	156	61 07	57	156	62 04	57	155	63 01	56	154	63 57	56	153
13	59 52	57	154	60 49	56	154	61 45	56	153	62 41	56	152	63 37	55	151
14	59 32	56	153	60 28	56	152	61 24	56	151	62 20	55	150	63 15	55	150
15	59 11	+56	151	60 07	+55	150	61 02	+56	149	61 58	+54	148	62 52	+55	148
16	58 49	56	149	59 45	54	148	60 39	55	148	61 34	54	147	62 28	54	146
17	58 26	55	147	59 21	54	147	60 15	54	146	61 09	54	145	62 03	53	144
18	58 02	54	146	58 56	54	145	59 50	53	144	60 43	53	143	61 36	53	142
19	57 37	53	144	58 30	54	143	59 24	53	142	60 17	52	141	61 09	52	140
20	57 11	+53	143	58 04	+52	142	58 56	+53	141	59 49	+51	140	60 40	+52	139
21	56 43	53	141	57 36	52	140	58 28	52	139	59 20	51	138	60 11	51	137
22	56 15	52	139	57 07	52	138	57 59	51	138	58 50	51	137	59 41	50	135
23	55 46	52	138	56 38	51	137	57 29	50	136	58 19	50	135	59 09	50	134
24	55 16	51	136	56 07	51	136	56 58	50	135	57 48	49	134	58 37	49	132
25	54 46	+50	135	55 36	+50	134	56 26	+50	133	57 16	+49	132	58 05	+48	131
26	54 14	50	134	55 04	50	133	55 54	49	132	56 43	48	131	57 31	48	130
27	53 42	50	132	54 32	49	131	55 21	48	130	56 09	48	129	56 57	47	128
28	53 09	49	131	53 58	49	130	54 47	48	129	55 35	47	128	56 22	47	127
29	52 36	48	130	53 24	48	129	54 12	48	128	55 00	47	127	55 47	46	125
30	52 02	+48	128	52 50	+47	127	53 37	+47	126	54 24	+47	125	55 11	+45	124
31	51 27	48	127	52 15	47	126	53 02	46	125	53 48	46	124	54 34	45	123
32	50 52	47	126	51 39	46	125	52 25	46	124	53 11	46	123	53 57	45	122
33	50 16	46	125	51 02	47	124	51 49	45	123	52 34	45	122	53 19	45	120
34	49 39	47	124	50 26	45	123	51 11	46	121	51 57	44	120	52 41	44	119
35	49 03	+45	122	49 48	+46	121	50 34	+45	120	51 19	+44	119	52 03	+43	118
36	48 25	46	121	49 11	45	120	49 56	44	119	50 40	44	118	51 24	43	117
37	47 47	46	120	48 33	44	119	49 17	44	118	50 01	44	117	50 45	42	116
38	47 09	45	119	47 54	44	118	48 38	44	117	49 22	43	116	50 05	42	115
39	46 31	44	118	47 15	44	117	47 59	43	116	48 42	43	115	49 25	42	114
40	45 52	+44	117	46 36	+43	116	47 19	+43	115	48 02	+43	114	48 45	+41	113
41	45 13	43	116	45 56	43	115	46 39	43	114	47 22	42	113	48 04	42	112
42	44 33	43	115	45 16	43	114	45 59	42	113	46 41	42	112	47 23	41	111
43	43 53	43	114	44 36	43	113	45 19	42	112	46 01	41	111	46 42	41	110
44	43 13	42	113	43 55	43	112	44 38	41	111	45 19	42	110	46 01	40	109
45	42 32	+43	112	43 15	+42	111	43 57	+41	110	44 38	+41	109	45 19	+40	108
46	41 51	43	111	42 34	41	110	43 15	42	109	43 57	40	108	44 37	40	107
47	41 10	42	110	41 52	42	109	42 34	41	108	43 15	40	107	43 55	40	106
48	40 29	42	109	41 11	41	108	41 52	41	107	42 33	40	106	43 13	40	105
49	39 47	42	108	40 29	41	108	41 10	41	107	41 51	40	106	42 31	39	105
50	39 06	+41	108	39 47	+41	107	40 28	+40	106	41 08	+40	105	41 48	+39	104
51	38 24	41	107	39 05	41	106	39 46	40	105	40 26	39	104	41 05	39	103
52	37 42	41	106	38 23	40	105	39 03	40	104	39 43	39	103	40 22	39	102
53	36 59	41	105	37 40	40	104	38 20	40	103	39 00	40	102	39 40	38	101
54	36 17	40	104	36 57	41	103	37 38	39	102	38 17	39	101	38 56	39	100
55	35 34	+41	103	36 15	+40	103	36 55	+39	102	37 34	+39	101	38 13	+39	100
56	34 52	40	103	35 32	40	102	36 12	39	101	36 51	39	100	37 30	38	99
57	34 09	40	102	34 49	39	101	35 28	40	100	36 08	38	99	36 46	39	98
58	33 26	40	101	34 06	39	100	34 45	39	99	35 24	39	98	36 03	38	97
59	32 43	39	100	33 22	40	99	34 02	39	99	34 41	38	98	35 19	39	97
60	31 59	+40	100	32 39	+39	99	33 18	+39	98	33 57	+39	97	34 36	+38	96
61	31 16	40	99	31 56	39	98	32 35	39	97	33 14	38	96	33 52	38	95
62	30 33	39	98	31 12	39	97	31 51	39	96	32 30	38	95	33 08	38	94
63	29 49	39	97	30 28	40	96	31 08	38	96	31 46	39	95	32 25	38	94
64	29 05	40	97	29 45	39	96	30 24	39	95	31 03	38	94	31 41	38	93
65	28 22	+39	96	29 01	+39	95	29 40	+39	94	30 19	+38	93	30 57	+38	92
66	27 38	39	95	28 17	39	94	28 56	39	93	29 35	38	93	30 13	38	92
67	26 54	40	94	27 34	39	94	28 13	38	93	28 51	38	92	29 29	38	91
68	26 11	39	94	26 50	39	93	27 29	38	92	28 07	38	91	28 45	38	90
69	25 27	39	93	26 06	39	92	26 45	38	91	27 23	39	90	28 02	37	90

LHA	20° Hc	20° d	20° Z	21° Hc	21° d	21° Z	22° Hc	22° d	22° Z	23° Hc	23° d	23° Z	24° Hc	24° d	24° Z
	° ′	′	°	° ′	′	°	° ′	′	°	° ′	′	°	° ′	′	°
0	67 00	+60	180	68 00	+60	180	69 00	+60	180	70 00	+60	180	71 00	+60	180
1	66 59	60	178	67 59	60	178	68 59	60	177	69 59	60	177	70 59	60	177
2	66 56	60	175	67 56	60	175	68 56	60	175	69 56	60	175	70 56	60	174
3	66 52	59	173	67 51	60	173	68 51	60	172	69 51	59	172	70 50	60	172
4	66 45	60	170	67 45	59	170	68 44	60	170	69 44	59	169	70 43	59	169
5	66 37	+59	168	67 36	+60	168	68 36	+59	167	69 35	+59	167	70 34	+58	166
6	66 27	59	166	67 26	59	165	68 25	59	165	69 24	58	164	70 22	58	164
7	66 16	58	164	67 14	58	163	68 12	59	162	69 11	58	162	70 09	57	161
8	66 02	58	161	67 00	58	161	67 58	58	160	68 56	57	159	69 53	57	158
9	65 47	58	159	66 45	57	158	67 42	57	158	68 39	57	157	69 36	57	156
10	65 31	+57	157	66 28	+57	156	67 25	+56	155	68 21	+56	154	69 17	+56	153
11	65 13	56	155	66 09	56	154	67 05	56	153	68 01	56	152	68 57	55	151
12	64 53	56	153	65 49	56	152	66 45	55	151	67 40	55	150	68 35	54	149
13	64 32	56	151	65 28	55	150	66 23	54	149	67 17	54	148	68 11	54	146
14	64 10	55	149	65 05	54	148	65 59	54	147	66 53	53	145	67 46	53	144
15	63 47	+54	147	64 41	+53	146	65 34	+53	145	66 27	+53	143	67 20	+52	142
16	63 22	53	145	64 15	53	144	65 08	53	143	66 01	51	141	66 52	52	140
17	62 56	53	143	63 49	52	142	64 41	52	141	65 33	51	140	66 24	50	138
18	62 29	52	141	63 21	52	140	64 13	51	139	65 04	50	138	65 54	50	136
19	62 01	51	139	62 52	51	138	63 43	50	137	64 33	50	136	65 23	49	134
20	61 32	+51	138	62 23	+50	137	63 13	+49	135	64 02	+49	134	64 51	+48	133
21	61 02	50	136	61 52	49	135	62 41	49	134	63 30	49	132	64 19	47	131
22	60 31	49	134	61 20	49	133	62 09	49	132	62 58	47	131	63 45	47	129
23	59 59	49	133	60 48	48	132	61 36	48	130	62 24	47	129	63 11	46	128
24	59 26	49	131	60 15	47	130	61 02	47	129	61 49	47	128	62 36	45	126
25	58 53	+48	130	59 41	+47	129	60 28	+46	127	61 14	+46	126	62 00	+45	125
26	58 19	47	128	59 06	47	127	59 53	45	126	60 38	45	125	61 23	45	123
27	57 44	47	127	58 31	46	126	59 17	45	125	60 02	44	123	60 46	44	122
28	57 09	46	126	57 55	45	124	58 40	45	123	59 25	44	122	60 09	43	121
29	56 33	45	124	57 18	45	123	58 03	45	122	58 48	43	121	59 31	42	119
30	55 56	+45	123	56 41	+45	122	57 26	+43	121	58 09	+43	119	58 52	+42	118
31	55 19	45	122	56 04	44	121	56 48	43	119	57 31	42	118	58 13	42	117
32	54 42	44	121	55 26	43	119	56 09	43	118	56 52	42	117	57 34	41	116
33	54 04	43	119	54 47	43	118	55 30	43	117	56 13	41	116	56 54	41	114
34	53 25	43	118	54 08	43	117	54 51	42	116	55 33	41	115	56 14	40	113
35	52 46	+43	117	53 29	+42	116	54 11	+42	115	54 53	+40	113	55 33	+40	112
36	52 07	42	116	52 49	42	115	53 31	41	114	54 12	41	112	54 53	39	111
37	51 27	42	115	52 09	42	114	52 51	41	113	53 32	40	111	54 12	39	110
38	50 47	42	114	51 29	41	113	52 10	41	112	52 51	39	110	53 30	39	109
39	50 07	41	113	50 48	41	112	51 29	40	110	52 09	40	109	52 49	38	108
40	49 26	+42	112	50 08	+40	111	50 48	+40	109	51 28	+39	108	52 07	+38	107
41	48 46	40	111	49 26	40	110	50 06	40	109	50 46	39	107	51 25	38	106
42	48 04	41	110	48 45	40	109	49 25	39	108	50 04	38	106	50 42	38	105
43	47 23	40	109	48 03	40	108	48 43	39	107	49 22	38	105	50 00	37	104
44	46 41	40	108	47 21	40	107	48 01	38	106	48 39	38	105	49 17	38	103
45	45 59	+40	107	46 39	+39	106	47 18	+39	105	47 57	+38	104	48 35	+37	103
46	45 17	40	106	45 57	39	105	46 36	38	104	47 14	38	103	47 52	37	102
47	44 35	39	105	45 14	39	104	45 53	38	103	46 31	38	102	47 09	36	101
48	43 53	39	104	44 32	38	103	45 10	38	102	45 48	37	101	46 25	37	100
49	43 10	39	104	43 49	38	103	44 27	38	101	45 05	37	100	45 42	37	99
50	42 27	+39	103	43 06	+38	102	43 44	+38	101	44 22	+37	100	44 59	+36	98
51	41 44	39	102	42 23	38	101	43 01	38	100	43 39	36	99	44 15	37	98
52	41 01	39	101	41 40	38	100	42 18	37	99	42 55	37	98	43 32	36	97
53	40 18	39	100	40 57	37	99	41 34	38	98	42 12	36	97	42 48	36	96
54	39 35	38	100	40 13	38	98	40 51	37	97	41 28	37	96	42 05	35	95
55	38 52	+38	99	39 30	+37	98	40 07	+37	97	40 44	+37	96	41 21	+36	95
56	38 08	38	98	38 46	38	97	39 24	37	96	40 01	36	95	40 37	36	94
57	37 25	38	97	38 03	37	96	38 40	37	95	39 17	36	94	39 53	36	93
58	36 41	38	96	37 19	37	95	37 56	37	94	38 33	36	93	39 09	36	92
59	35 58	37	96	36 35	38	95	37 13	36	94	37 49	37	93	38 26	35	92
60	35 14	+38	95	35 52	+37	94	36 29	+37	93	37 06	+36	92	37 42	+35	91
61	34 30	38	94	35 08	37	93	35 45	37	92	36 22	36	91	36 58	36	90
62	33 46	38	94	34 24	37	93	35 01	37	92	35 38	36	91	36 14	36	90
63	33 03	37	93	33 40	37	92	34 17	37	91	34 54	36	90	35 30	36	89
64	32 19	37	92	32 56	37	91	33 33	37	90	34 10	36	89	34 46	36	88
65	31 35	+37	91	32 12	+37	91	32 49	+37	90	33 26	+36	89	34 02	+36	88
66	30 51	38	91	31 29	37	90	32 06	36	89	32 42	37	88	33 19	35	87
67	30 07	38	90	30 45	37	89	31 22	36	88	31 58	37	87	32 35	36	86
68	29 23	38	89	30 01	37	88	30 38	37	88	31 15	36	87	31 51	36	86
69	28 39	38	89	29 17	37	88	29 54	37	87	30 31	36	86	31 07	36	85

LHA	25° Hc	25° d	25° Z	26° Hc	26° d	26° Z	27° Hc	27° d	27° Z	28° Hc	28° d	28° Z	29° Hc	29° d	29° Z	LHA
	° ′	′	°	° ′	′	°	° ′	′	°	° ′	′	°	° ′	′	°	
0	72 00	+60	180	73 00	+60	180	74 00	+60	180	75 00	+60	180	76 00	+60	180	360
1	71 59	60	177	72 59	60	177	73 59	60	177	74 59	60	177	75 59	60	176	359
2	71 56	59	174	72 55	60	174	73 55	60	174	74 55	60	173	75 55	59	173	358
3	71 50	60	171	72 50	59	171	73 49	59	170	74 48	60	170	75 48	59	169	357
4	71 42	59	168	72 41	59	168	73 40	59	167	74 39	59	167	75 38	59	166	356
5	71 32	+59	166	72 31	+59	165	73 30	+58	164	74 28	+58	163	75 26	+58	162	355
6	71 20	59	163	72 19	57	162	73 16	58	161	74 14	58	160	75 12	57	159	354
7	71 06	58	160	72 04	57	159	73 01	57	158	73 58	57	157	74 55	56	156	353
8	70 50	57	157	71 47	57	156	72 44	56	155	73 40	56	154	74 36	55	153	352
9	70 33	56	155	71 29	56	154	72 25	55	153	73 20	55	151	74 15	54	150	351
10	70 13	+56	152	71 09	+54	151	72 03	+55	150	72 58	+54	148	73 52	+53	147	350
11	69 52	54	150	70 46	55	149	71 41	53	147	72 34	53	146	73 27	52	144	349
12	69 29	54	148	70 23	53	146	71 16	53	145	72 09	51	143	73 00	51	142	348
13	69 05	53	145	69 58	52	144	70 50	52	142	71 42	50	141	72 32	50	139	347
14	68 39	52	143	69 31	51	142	70 22	51	140	71 13	50	138	72 03	48	137	346
15	68 12	+51	141	69 03	+51	139	69 54	+49	138	70 43	+49	136	71 32	+47	134	345
16	67 44	50	139	68 34	50	137	69 24	48	136	70 12	48	134	71 00	47	132	344
17	67 14	50	137	68 04	48	135	68 52	48	134	69 40	47	132	70 27	46	130	343
18	66 44	48	135	67 32	48	133	68 20	47	132	69 07	46	130	69 53	45	128	342
19	66 12	48	133	67 00	47	132	67 47	46	130	68 33	45	128	69 18	44	126	341
20	65 39	+48	131	66 27	+46	130	67 13	+45	128	67 58	+44	126	68 42	+43	125	340
21	65 06	46	130	65 52	46	128	66 38	44	126	67 22	44	125	68 06	42	123	339
22	64 32	45	128	65 17	45	126	66 02	44	125	66 46	43	123	67 29	41	121	338
23	63 57	45	126	64 42	44	125	65 26	43	123	66 09	42	121	66 51	40	120	337
24	63 21	44	125	64 05	44	123	64 49	42	122	65 31	41	120	66 12	40	118	336
25	62 45	+43	123	63 28	+43	122	64 11	+42	120	64 53	+40	119	65 33	+40	117	335
26	62 08	43	122	62 51	42	120	63 33	41	119	64 14	40	117	64 54	39	115	334
27	61 30	43	120	62 13	41	119	62 54	41	117	63 35	39	116	64 14	38	114	333
28	60 52	42	119	61 34	41	118	62 15	40	116	62 55	39	114	63 34	37	113	332
29	60 13	42	118	60 55	40	116	61 35	40	115	62 15	38	113	62 53	37	112	331
30	59 34	+41	117	60 15	+40	115	60 55	+39	114	61 34	+38	112	62 12	+37	110	330
31	58 55	40	115	59 35	40	114	60 15	38	112	60 53	38	111	61 31	36	109	329
32	58 15	40	114	58 55	39	113	59 34	38	111	60 12	37	110	60 49	36	108	328
33	57 35	39	113	58 14	39	112	58 53	38	110	59 31	36	109	60 07	36	107	327
34	56 54	39	112	57 33	39	111	58 12	37	109	58 49	36	108	59 25	35	106	326
35	56 13	+39	111	56 52	+38	109	57 30	+37	108	58 07	+36	107	58 43	+35	105	325
36	55 32	39	110	56 11	37	108	56 48	37	107	57 25	36	106	58 01	34	104	324
37	54 51	38	109	55 29	37	107	56 06	37	106	56 43	35	105	57 18	34	103	323
38	54 09	38	108	54 47	37	106	55 24	36	105	56 00	35	104	56 35	34	102	322
39	53 27	38	107	54 05	36	105	54 41	36	104	55 17	35	103	55 52	34	101	321
40	52 45	+37	106	53 22	+37	105	53 59	+35	103	54 34	+35	102	55 09	+34	100	320
41	52 03	37	105	52 40	36	104	53 16	35	102	53 51	35	101	54 26	33	100	319
42	51 20	37	104	51 57	36	103	52 33	35	101	53 08	34	100	53 42	34	99	318
43	50 37	37	103	51 14	36	102	51 50	35	101	52 25	34	99	52 59	33	98	317
44	49 55	36	102	50 31	36	101	51 07	35	100	51 42	33	98	52 15	33	97	316
45	49 12	+36	101	49 48	+35	100	50 23	+35	99	50 58	+34	98	51 32	+33	96	315
46	48 29	36	101	49 05	35	99	49 40	35	98	50 15	33	97	50 48	33	95	314
47	47 45	36	100	48 21	36	98	48 57	34	97	49 31	34	96	50 05	32	95	313
48	47 02	36	99	47 38	35	98	48 13	34	96	48 47	34	95	49 21	32	94	312
49	46 19	35	98	46 54	35	97	47 29	35	96	48 04	33	94	48 37	33	93	311
50	45 35	+36	97	46 11	+35	96	46 46	+34	95	47 20	+33	94	47 53	+33	93	310
51	44 52	35	97	45 27	35	95	46 02	34	94	46 36	33	93	47 09	33	92	309
52	44 08	35	96	44 43	35	95	45 18	34	93	45 52	33	92	46 25	33	91	308
53	43 24	36	95	44 00	34	94	44 34	34	93	45 08	34	92	45 42	32	90	307
54	42 40	36	94	43 16	34	93	43 50	34	92	44 24	34	91	44 58	32	90	306
55	41 57	+35	94	42 32	+35	92	43 07	+34	91	43 41	+33	90	44 14	+32	89	305
56	41 13	35	93	41 48	35	92	42 23	34	91	42 57	33	90	43 30	33	88	304
57	40 29	35	92	41 04	35	91	41 39	34	90	42 13	33	89	42 46	33	88	303
58	39 45	35	91	40 20	35	90	40 55	34	89	41 29	33	88	42 02	33	87	302
59	39 01	35	91	39 36	35	90	40 11	34	89	40 45	33	88	41 18	33	86	301
60	38 17	+36	90	38 53	+34	89	39 27	+34	88	40 01	+34	87	40 35	+32	86	300
61	37 34	35	89	38 09	34	88	38 43	34	87	39 17	34	86	39 51	33	85	299
62	36 50	35	89	37 25	35	88	38 00	34	87	38 34	33	86	39 07	33	85	298
63	36 06	35	88	36 41	35	87	37 16	34	86	37 50	34	85	38 24	33	84	297
64	35 22	35	87	35 57	35	86	36 32	34	85	37 06	34	84	37 40	33	83	296
65	34 38	+35	87	35 13	+35	86	35 48	+35	85	36 23	+33	84	36 56	+34	83	295
66	33 54	36	86	34 30	35	85	35 05	34	84	35 39	34	83	36 13	33	82	294
67	33 11	35	85	33 46	35	84	34 21	35	83	34 56	34	82	35 30	33	81	293
68	32 27	35	85	33 02	35	84	33 37	35	83	34 12	34	82	34 46	34	81	292
69	31 43	36	84	32 19	35	83	32 54	35	82	33 29	34	81	34 03	34	80	291

S. Lat. {LHA greater than 180°....... Zn=180−Z
{LHA less than 180°.......... Zn=180+Z

DECLINATION (15°-29°) SAME NAME AS LATITUDE

LAT 43°

TABLE 5.—Correction to Tabulated Altitude for Minutes of Declination

d / ′	1	2	3	4	5	6	7	8	9	10	11	12	13	14	15	16	17	18	19	20
0	0	0	0	0	0	0	0	0	0	0	0	0	0	0	0	0	0	0	0	0
1	0	0	0	0	0	0	0	0	0	0	0	0	0	0	0	0	0	0	0	0
2	0	0	0	0	0	0	0	0	0	0	0	0	0	0	0	1	1	1	1	1
3	0	0	0	0	0	0	0	0	0	0	1	1	1	1	1	1	1	1	1	1
4	0	0	0	0	0	0	0	1	1	1	1	1	1	1	1	1	1	1	1	1
5	0	0	0	0	0	0	1	1	1	1	1	1	1	1	1	1	1	2	2	2
6	0	0	0	0	0	1	1	1	1	1	1	1	1	1	2	2	2	2	2	2
7	0	0	0	0	1	1	1	1	1	1	1	1	2	2	2	2	2	2	2	2
8	0	0	0	1	1	1	1	1	1	1	1	2	2	2	2	2	2	2	3	3
9	0	0	0	1	1	1	1	1	1	2	2	2	2	2	2	2	3	3	3	3
10	0	0	0	1	1	1	1	1	2	2	2	2	2	2	2	3	3	3	3	3
11	0	0	1	1	1	1	1	1	2	2	2	2	2	3	3	3	3	3	3	4
12	0	0	1	1	1	1	1	2	2	2	2	2	3	3	3	3	3	4	4	4
13	0	0	1	1	1	1	2	2	2	2	2	3	3	3	3	3	4	4	4	4
14	0	0	1	1	1	1	2	2	2	2	3	3	3	3	4	4	4	4	4	5
15	0	0	1	1	1	2	2	2	2	2	3	3	3	4	4	4	4	4	5	5
16	0	1	1	1	1	2	2	2	2	3	3	3	3	4	4	4	5	5	5	5
17	0	1	1	1	1	2	2	2	3	3	3	3	4	4	4	5	5	5	5	6
18	0	1	1	1	2	2	2	2	3	3	3	4	4	4	4	5	5	5	6	6
19	0	1	1	1	2	2	2	3	3	3	3	4	4	4	5	5	5	6	6	6
20	0	1	1	1	2	2	2	3	3	3	4	4	4	5	5	5	6	6	6	7
21	0	1	1	1	2	2	2	3	3	4	4	4	5	5	5	6	6	6	7	7
22	0	1	1	1	2	2	3	3	3	4	4	4	5	5	6	6	6	7	7	7
23	0	1	1	2	2	2	3	3	3	4	4	5	5	5	6	6	7	7	7	8
24	0	1	1	2	2	2	3	3	4	4	4	5	5	6	6	6	7	7	8	8
25	0	1	1	2	2	2	3	3	4	4	5	5	5	6	6	7	7	8	8	8
26	0	1	1	2	2	3	3	3	4	4	5	5	6	6	6	7	7	8	8	9
27	0	1	1	2	2	3	3	4	4	4	5	5	6	6	7	7	8	8	9	9
28	0	1	1	2	2	3	3	4	4	5	5	6	6	7	7	7	8	8	9	9
29	0	1	1	2	2	3	3	4	4	5	5	6	6	7	7	8	8	9	9	10
30	0	1	2	2	2	3	4	4	4	5	6	6	6	7	8	8	8	9	10	10
31	1	1	2	2	3	3	4	4	5	5	6	6	7	7	8	8	9	9	10	10
32	1	1	2	2	3	3	4	4	5	5	6	6	7	7	8	9	9	10	10	11
33	1	1	2	2	3	3	4	4	5	6	6	7	7	8	8	9	9	10	10	11
34	1	1	2	2	3	3	4	5	5	6	6	7	7	8	8	9	10	10	11	11
35	1	1	2	2	3	4	4	5	5	6	6	7	8	8	9	9	10	10	11	12
36	1	1	2	2	3	4	4	5	5	6	7	7	8	8	9	10	10	11	11	12
37	1	1	2	2	3	4	4	5	6	6	7	7	8	9	9	10	10	11	12	12
38	1	1	2	3	3	4	4	5	6	6	7	8	8	9	10	10	11	11	12	13
39	1	1	2	3	3	4	5	5	6	6	7	8	8	9	10	10	11	12	12	13
40	1	1	2	3	3	4	5	5	6	7	7	8	9	9	10	11	11	12	13	13
41	1	1	2	3	3	4	5	5	6	7	8	8	9	10	10	11	12	12	13	14
42	1	1	2	3	4	4	5	6	6	7	8	8	9	10	10	11	12	13	13	14
43	1	1	2	3	4	4	5	6	6	7	8	9	9	10	11	11	12	13	14	14
44	1	1	2	3	4	4	5	6	7	7	8	9	10	10	11	12	12	13	14	15
45	1	2	2	3	4	4	5	6	7	8	8	9	10	10	11	12	13	14	14	15
46	1	2	2	3	4	5	5	6	7	8	8	9	10	11	12	12	13	14	15	15
47	1	2	2	3	4	5	5	6	7	8	9	9	10	11	12	13	13	14	15	16
48	1	2	2	3	4	5	6	6	7	8	9	10	10	11	12	13	14	14	15	16
49	1	2	2	3	4	5	6	7	7	8	9	10	11	11	12	13	14	15	16	16
50	1	2	2	3	4	5	6	7	8	8	9	10	11	12	12	13	14	15	16	17
51	1	2	3	3	4	5	6	7	8	8	9	10	11	12	13	14	14	15	16	17
52	1	2	3	3	4	5	6	7	8	9	10	10	11	12	13	14	15	16	16	17
53	1	2	3	4	4	5	6	7	8	9	10	11	11	12	13	14	15	16	17	18
54	1	2	3	4	4	5	6	7	8	9	10	11	12	13	14	14	15	16	17	18
55	1	2	3	4	5	6	6	7	8	9	10	11	12	13	14	15	16	16	17	18
56	1	2	3	4	5	6	7	7	8	9	10	11	12	13	14	15	16	17	18	19
57	1	2	3	4	5	6	7	8	9	10	10	11	12	13	14	15	16	17	18	19
58	1	2	3	4	5	6	7	8	9	10	11	12	13	14	14	15	16	17	18	19
59	1	2	3	4	5	6	7	8	9	10	11	12	13	14	15	16	17	18	19	20

d / ′	21	22	23	24	25	26	27	28	29	30	31	32	33	34	35	36	37	38	39	40
0	0	0	0	0	0	0	0	0	0	0	0	0	0	0	0	0	0	0	0	0
1	0	0	0	0	0	0	0	0	0	0	1	1	1	1	1	1	1	1	1	1
2	1	1	1	1	1	1	1	1	1	1	1	1	1	1	1	1	1	1	1	1
3	1	1	1	1	1	1	1	1	1	2	2	2	2	2	2	2	2	2	2	2
4	1	1	2	2	2	2	2	2	2	2	2	2	2	2	2	2	2	3	3	3
5	2	2	2	2	2	2	2	2	2	2	3	3	3	3	3	3	3	3	3	3
6	2	2	2	2	2	3	3	3	3	3	3	3	3	3	4	4	4	4	4	4
7	2	3	3	3	3	3	3	3	3	4	4	4	4	4	4	4	4	4	5	5
8	3	3	3	3	3	3	4	4	4	4	4	4	4	5	5	5	5	5	5	5
9	3	3	3	4	4	4	4	4	4	4	5	5	5	5	5	5	6	6	6	6
10	4	4	4	4	4	4	4	5	5	5	5	5	6	6	6	6	6	6	6	7
11	4	4	4	4	5	5	5	5	5	6	6	6	6	6	6	7	7	7	7	7
12	4	4	5	5	5	5	5	6	6	6	6	6	7	7	7	7	7	8	8	8
13	5	5	5	5	5	6	6	6	6	6	7	7	7	7	8	8	8	8	8	9
14	5	5	5	6	6	6	6	7	7	7	7	7	8	8	8	8	9	9	9	9
15	5	6	6	6	6	6	7	7	7	8	8	8	8	8	9	9	9	10	10	10
16	6	6	6	6	7	7	7	7	8	8	8	9	9	9	9	10	10	10	10	11
17	6	6	7	7	7	7	8	8	8	8	9	9	9	10	10	10	10	11	11	11
18	6	7	7	7	8	8	8	8	9	9	9	10	10	10	10	11	11	11	12	12
19	7	7	7	8	8	8	9	9	9	10	10	10	10	11	11	11	12	12	12	13
20	7	7	8	8	8	9	9	9	10	10	10	11	11	11	12	12	12	13	13	13
21	7	8	8	8	9	9	9	10	10	10	11	11	12	12	12	13	13	13	14	14
22	8	8	8	9	9	10	10	10	11	11	11	12	12	12	13	13	14	14	14	15
23	8	8	9	9	10	10	10	11	11	12	12	12	13	13	13	14	14	15	15	15
24	8	9	9	10	10	10	11	11	12	12	12	13	13	14	14	14	15	15	16	16
25	9	9	10	10	10	11	11	12	12	12	13	13	14	14	15	15	15	16	16	17
26	9	10	10	10	11	11	12	12	13	13	13	14	14	15	15	16	16	16	17	17
27	9	10	10	11	11	12	12	13	13	14	14	14	15	15	16	16	17	17	18	18
28	10	10	11	11	12	12	13	13	14	14	14	15	15	16	16	17	17	18	18	19
29	10	11	11	12	12	13	13	14	14	14	15	15	16	16	17	17	18	18	19	19
30	10	11	12	12	12	13	14	14	14	15	16	16	16	17	18	18	18	19	20	20
31	11	11	12	12	13	13	14	14	15	16	16	17	17	18	18	19	19	20	20	21
32	11	12	12	13	13	14	14	15	15	16	17	17	18	18	19	19	20	20	21	21
33	12	12	13	13	14	14	15	15	16	16	17	18	18	19	19	20	20	21	21	22
34	12	12	13	14	14	15	15	16	16	17	18	18	19	19	20	20	21	22	22	23
35	12	13	13	14	15	15	16	16	17	18	18	19	19	20	20	21	22	22	23	23
36	13	13	14	14	15	16	16	17	17	18	19	19	20	20	21	22	22	23	23	24
37	13	14	14	15	15	16	17	17	18	18	19	20	20	21	22	22	23	23	24	25
38	13	14	15	15	16	16	17	18	18	19	20	20	21	22	22	23	23	24	25	25
39	14	14	15	16	16	17	18	18	19	20	20	21	21	22	23	23	24	25	25	26
40	14	15	15	16	17	17	18	19	19	20	21	21	22	23	23	24	25	25	26	27
41	14	15	16	16	17	18	18	19	20	20	21	22	23	23	24	25	25	26	27	27
42	15	15	16	17	18	18	19	20	20	21	22	22	23	24	24	25	26	27	27	28
43	15	16	16	17	18	19	19	20	21	22	22	23	24	24	25	26	27	27	28	29
44	15	16	17	18	18	19	20	21	21	22	23	23	24	25	26	26	27	28	29	29
45	16	16	17	18	19	20	20	21	22	22	23	24	25	26	26	27	28	28	29	30
46	16	17	18	18	19	20	21	21	22	23	24	25	25	26	27	28	28	29	30	31
47	16	17	18	19	20	20	21	22	23	24	24	25	26	27	27	28	29	30	31	31
48	17	18	18	19	20	21	22	22	23	24	25	26	26	27	28	29	30	30	31	32
49	17	18	19	20	20	21	22	23	24	24	25	26	27	28	29	29	30	31	32	33
50	18	18	19	20	21	22	22	23	24	25	26	27	28	28	29	30	31	32	32	33
51	18	19	20	20	21	22	23	24	25	26	26	27	28	29	30	31	31	32	33	34
52	18	19	20	21	22	23	23	24	25	26	27	28	29	29	30	31	32	33	34	35
53	19	19	20	21	22	23	24	25	26	26	27	28	29	30	31	32	33	34	34	35
54	19	20	21	22	22	23	24	25	26	27	28	29	30	31	32	32	33	34	35	36
55	19	20	21	22	23	24	25	26	27	28	28	29	30	31	32	33	34	35	36	37
56	20	21	21	22	23	24	25	26	27	28	29	30	31	32	33	34	35	35	36	37
57	20	21	22	23	24	25	26	27	28	28	29	30	31	32	33	34	35	36	37	38
58	20	21	22	23	24	25	26	27	28	29	30	31	32	33	34	35	36	37	38	39
59	21	22	23	24	25	26	27	28	29	30	30	31	32	33	34	35	36	37	38	39

d / ′	41	42	43	44	45	46	47	48	49	50	51	52	53	54	55	56	57	58	59	60	d / ′
0	0	0	0	0	0	0	0	0	0	0	0	0	0	0	0	0	0	0	0	0	0
1	1	1	1	1	1	1	1	1	1	1	1	1	1	1	1	1	1	1	1	1	1
2	1	1	1	1	2	2	2	2	2	2	2	2	2	2	2	2	2	2	2	2	2
3	2	2	2	2	2	2	2	2	2	2	3	3	3	3	3	3	3	3	3	3	3
4	3	3	3	3	3	3	3	3	3	3	3	3	4	4	4	4	4	4	4	4	4
5	3	4	4	4	4	4	4	4	4	4	4	4	4	4	5	5	5	5	5	5	5
6	4	4	4	4	4	5	5	5	5	5	5	5	5	5	6	6	6	6	6	6	6
7	5	5	5	5	5	5	5	6	6	6	6	6	6	6	6	7	7	7	7	7	7
8	5	6	6	6	6	6	6	6	7	7	7	7	7	7	7	7	8	8	8	8	8
9	6	6	6	7	7	7	7	7	7	8	8	8	8	8	8	8	9	9	9	9	9
10	7	7	7	7	8	8	8	8	8	8	8	9	9	9	9	9	10	10	10	10	10
11	8	8	8	8	8	8	9	9	9	9	9	10	10	10	10	10	10	11	11	11	11
12	8	8	9	9	9	9	9	10	10	10	10	10	11	11	11	11	11	12	12	12	12
13	9	9	9	10	10	10	10	10	11	11	11	11	11	12	12	12	12	13	13	13	13
14	10	10	10	10	10	11	11	11	11	12	12	12	12	13	13	13	13	14	14	14	14
15	10	10	11	11	11	12	12	12	12	12	13	13	13	14	14	14	14	14	15	15	15
16	11	11	11	12	12	12	13	13	13	13	14	14	14	14	15	15	15	15	16	16	16
17	12	12	12	12	13	13	13	14	14	14	14	15	15	15	16	16	16	16	17	17	17
18	12	13	13	13	14	14	14	14	15	15	15	16	16	16	16	17	17	17	18	18	18
19	13	13	14	14	14	15	15	15	16	16	16	16	17	17	17	18	18	18	19	19	19
20	14	14	14	15	15	15	16	16	16	17	17	17	18	18	18	19	19	19	20	20	20
21	14	15	15	15	16	16	16	17	17	18	18	18	19	19	19	20	20	20	21	21	21
22	15	15	16	16	16	17	17	18	18	18	19	19	19	20	20	21	21	21	22	22	22
23	16	16	16	17	17	18	18	18	19	19	20	20	20	21	21	21	22	22	23	23	23
24	16	17	17	18	18	18	19	19	20	20	20	21	21	22	22	22	23	23	24	24	24
25	17	18	18	18	19	19	20	20	20	21	21	22	22	22	23	23	24	24	25	25	25
26	18	18	19	19	20	20	20	21	21	22	22	23	23	23	24	24	25	25	26	26	26
27	18	19	19	20	20	21	21	22	22	22	23	23	24	24	25	25	26	26	27	27	27
28	19	20	20	21	21	21	22	22	23	23	24	24	25	25	26	26	27	27	28	28	28
29	20	20	21	21	22	22	23	23	24	24	25	25	26	26	27	27	28	28	29	29	29
30	20	21	22	22	22	23	24	24	24	25	26	26	26	27	28	28	28	29	30	30	30
31	21	22	22	23	23	24	24	25	25	26	26	27	27	28	28	29	29	30	30	31	31
32	22	22	23	23	24	25	25	26	26	27	27	28	28	29	29	30	30	31	31	32	32
33	23	23	24	24	25	25	26	26	27	28	28	29	29	30	30	31	31	32	32	33	33
34	23	24	24	25	26	26	27	27	28	28	29	29	30	31	31	32	32	33	33	34	34
35	24	24	25	26	26	27	27	28	29	29	30	30	31	32	32	33	33	34	34	35	35
36	25	25	26	26	27	28	28	29	29	30	31	31	32	32	33	34	34	35	35	36	36
37	25	26	27	27	28	28	29	30	30	31	31	32	33	33	34	35	35	36	36	37	37
38	26	27	27	28	28	29	30	30	31	32	32	33	34	34	35	35	36	37	37	38	38
39	27	27	28	29	29	30	31	31	32	32	33	34	34	35	36	36	37	38	38	39	39
40	27	28	29	29	30	31	31	32	33	33	34	35	35	36	37	37	38	39	39	40	40
41	28	29	29	30	31	31	32	33	33	34	35	36	36	37	38	38	39	40	40	41	41
42	29	29	30	31	32	32	33	34	34	35	36	36	37	38	38	39	40	41	41	42	42
43	29	30	31	32	32	33	34	34	35	36	37	37	38	39	39	40	41	42	42	43	43
44	30	31	32	32	33	34	34	35	36	37	37	38	39	40	40	41	42	43	43	44	44
45	31	32	32	33	34	34	35	36	37	38	38	39	40	40	41	42	43	44	44	45	45
46	31	32	33	34	34	35	36	37	38	38	39	40	41	41	42	43	44	44	45	46	46
47	32	33	34	34	35	36	37	38	38	39	40	41	42	42	43	44	45	45	46	47	47
48	33	34	34	35	36	37	38	38	39	40	41	42	42	43	44	45	46	46	47	48	48
49	33	34	35	36	37	38	38	39	40	41	42	42	43	44	45	46	47	47	48	49	49
50	34	35	36	37	38	38	39	40	41	42	42	43	44	45	46	47	48	48	49	50	50
51	35	36	37	37	38	39	40	41	42	42	43	44	45	46	47	48	48	49	50	51	51
52	36	36	37	38	39	40	41	42	42	43	44	45	46	47	48	49	49	50	51	52	52
53	36	37	38	39	40	41	42	42	43	44	45	46	47	48	49	49	50	51	52	53	53
54	37	38	39	40	40	41	42	43	44	45	46	47	48	49	50	50	51	52	53	54	54
55	38	38	39	40	41	42	43	44	45	46	47	48	49	50	50	51	52	53	54	55	55
56	38	39	40	41	42	43	44	45	46	47	48	49	49	50	51	52	53	54	55	56	56
57	39	40	41	42	43	44	45	46	47	48	48	49	50	51	52	53	54	55	56	57	57
58	40	41	42	43	44	44	45	46	47	48	49	50	51	52	53	54	55	56	57	58	58
59	40	41	42	43	44	45	46	47	48	49	50	51	52	53	54	55	56	57	58	59	59

 LAT 42°N

LHA ♈	Hc Zn	Hc Zn	Hc Zn	Hc Zn	Hc Zn	Hc Zn	Hc Zn
	Kochab	♦VEGA	ARCTURUS	♦SPICA	REGULUS	♦POLLUX	CAPELLA
180	52 24 018	19 08 055	53 35 118	33 32 155	51 18 228	37 03 277	22 23 313
181	52 38 017	19 44 055	54 14 119	33 50 156	50 45 229	36 19 277	21 51 313
182	52 51 017	20 21 056	54 52 120	34 08 157	50 11 230	35 35 278	21 18 313
183	53 04 017	20 58 056	55 31 122	34 25 158	49 37 231	34 51 278	20 46 314
184	53 17 016	21 35 057	56 08 123	34 41 160	49 02 233	34 07 279	20 14 314
185	53 29 016	22 12 057	56 45 124	34 56 161	48 26 234	33 23 280	19 42 315
186	53 42 016	22 50 058	57 22 126	35 10 162	47 50 235	32 39 280	19 11 315
187	53 54 016	23 27 058	57 58 127	35 24 163	47 13 236	31 55 281	18 39 316
188	54 06 015	24 05 059	58 33 128	35 36 164	46 36 237	31 11 281	18 08 316
189	54 17 015	24 44 059	59 08 130	35 48 165	45 58 238	30 27 282	17 38 317
190	54 29 015	25 22 060	59 42 131	35 59 167	45 20 239	29 44 282	17 07 317
191	54 40 014	26 01 060	60 15 133	36 09 168	44 41 240	29 00 283	16 37 318
192	54 51 014	26 39 061	60 47 134	36 18 169	44 02 241	28 17 284	16 08 318
193	55 01 014	27 18 061	61 19 136	36 26 170	43 23 242	27 34 284	15 38 319
194	55 12 013	27 57 062	61 49 138	36 33 171	42 43 243	26 50 285	15 09 319
	♦VEGA	Rasalhague	ARCTURUS	♦SPICA	REGULUS	♦POLLUX	Dubhe
195	28 37 062	24 18 095	62 19 139	36 39 173	42 03 244	26 07 285	63 34 329
196	29 16 063	25 02 096	62 47 141	36 44 174	41 23 245	25 24 286	63 11 328
197	29 56 063	25 47 096	63 14 143	36 49 175	40 42 246	24 42 286	62 47 328
198	30 36 063	26 31 097	63 41 145	36 52 176	40 01 247	23 59 287	62 23 327
199	31 16 064	27 15 098	64 06 147	36 55 178	39 20 248	23 16 288	61 58 326
200	31 56 064	27 59 098	64 30 149	36 56 179	38 39 249	22 34 288	61 34 326
201	32 36 065	28 43 099	64 52 151	36 57 180	37 57 250	21 52 289	61 09 326
202	33 17 065	29 27 100	65 13 153	36 56 181	37 15 251	21 09 289	60 43 325
203	33 57 066	30 11 101	65 33 155	36 55 182	36 33 251	20 27 290	60 18 325
204	34 38 066	30 55 101	65 51 157	36 52 184	35 51 252	19 46 290	59 52 324
205	35 19 067	31 39 102	66 07 159	36 49 185	35 08 253	19 04 291	59 26 324
206	36 00 067	32 22 103	66 22 162	36 45 186	34 25 254	18 22 292	58 59 324
207	36 41 068	33 06 104	66 35 164	36 40 187	33 42 255	17 41 292	58 33 323
208	37 22 068	33 49 105	66 47 166	36 33 189	32 59 256	17 00 293	58 06 323
209	38 04 069	34 32 105	66 56 169	36 26 190	32 16 256	16 19 293	57 39 323
	♦DENEB	VEGA	Rasalhague	ANTARES	♦ARCTURUS	REGULUS	♦Dubhe
210	22 28 049	38 45 069	35 15 106	13 32 146	67 04 171	31 32 257	57 12 322
211	23 02 049	39 27 070	35 58 107	13 56 147	67 10 173	30 49 258	56 45 322
212	23 36 050	40 09 070	36 40 108	14 20 148	67 14 176	30 05 259	56 17 322
213	24 10 050	40 51 070	37 22 109	14 44 149	67 17 178	29 21 259	55 50 322
214	24 44 050	41 33 071	38 05 109	15 06 150	67 17 181	28 38 260	55 22 322
215	25 19 051	42 15 071	38 47 110	15 29 150	67 16 183	27 54 261	54 55 321
216	25 53 051	42 57 072	39 28 111	15 50 151	67 12 186	27 10 262	54 27 321
217	26 28 052	43 40 072	40 10 112	16 12 152	67 07 188	26 25 262	53 59 321
218	27 03 052	44 22 073	40 51 113	16 32 153	67 00 190	25 41 263	53 31 321
219	27 39 053	45 05 073	41 32 114	16 52 154	66 51 193	24 57 264	53 03 321
220	28 14 053	45 48 074	42 12 115	17 11 155	66 40 195	24 13 264	52 35 321
221	28 50 054	46 30 074	42 53 116	17 30 156	66 27 198	23 28 265	52 07 321
222	29 26 054	47 13 074	43 33 117	17 48 157	66 13 200	22 44 266	51 38 321
223	30 02 054	47 56 075	44 12 118	18 06 157	65 57 202	21 59 267	51 10 321
224	30 39 055	48 39 075	44 51 119	18 22 158	65 40 204	21 15 267	50 42 321
	DENEB	♦ALTAIR	Rasalhague	ANTARES	♦ARCTURUS	Denebola	♦Dubhe
225	31 15 055	18 54 095	45 30 120	18 38 159	65 21 206	40 36 251	50 14 321
226	31 52 056	19 39 096	46 09 121	18 54 160	65 00 208	39 53 252	49 45 321
227	32 29 056	20 23 097	46 47 122	19 09 161	64 38 210	39 11 253	49 17 321
228	33 06 056	21 07 097	47 24 123	19 23 162	64 15 212	38 28 254	48 49 321
229	33 43 057	21 51 098	48 01 124	19 36 163	63 50 214	37 45 255	48 20 321
230	34 20 057	22 35 099	48 38 125	19 49 164	63 24 216	37 02 255	47 52 321
231	34 58 058	23 19 099	49 14 127	20 01 165	62 57 218	36 19 256	47 24 321
232	35 36 058	24 03 100	49 50 128	20 13 166	62 29 220	35 36 257	46 56 321
233	36 14 058	24 47 101	50 25 129	20 23 167	62 00 222	34 52 258	46 28 321
234	36 52 059	25 31 102	50 59 130	20 33 168	61 30 223	34 09 259	45 59 321
235	37 30 059	26 15 102	51 33 132	20 43 168	60 59 225	33 25 259	45 31 321
236	38 08 060	26 58 103	52 06 133	20 51 169	60 27 227	32 41 260	45 03 321
237	38 47 060	27 41 104	52 38 134	20 59 170	59 54 228	31 57 261	44 35 321
238	39 25 060	28 25 105	53 10 136	21 06 171	59 21 230	31 13 262	44 07 321
239	40 04 061	29 08 105	53 41 137	21 12 172	58 46 231	30 29 262	43 39 321
	♦DENEB	ALTAIR	Rasalhague	♦ANTARES	ARCTURUS	Denebola	♦Dubhe
240	40 43 061	29 51 106	54 11 138	21 18 173	58 11 233	29 44 263	43 12 321
241	41 22 061	30 33 107	54 40 140	21 23 174	57 35 234	29 00 264	42 44 322
242	42 01 062	31 16 108	55 08 141	21 27 175	56 59 235	28 16 264	42 16 322
243	42 41 062	31 58 109	55 36 143	21 31 176	56 22 237	27 31 265	41 49 322
244	43 20 062	32 40 109	56 02 144	21 33 177	55 45 238	26 47 266	41 21 322
245	44 00 063	33 22 110	56 28 146	21 35 178	55 07 239	26 02 267	40 54 322
246	44 39 063	34 04 111	56 52 148	21 36 179	54 28 240	25 18 267	40 26 322
247	45 19 063	34 46 112	57 15 149	21 37 180	53 49 241	24 33 268	39 59 323
248	45 59 064	35 27 113	57 38 151	21 36 181	53 10 243	23 49 269	39 32 323
249	46 39 064	36 08 114	57 59 153	21 35 182	52 30 244	23 04 269	39 05 323
250	47 19 064	36 48 115	58 19 154	21 33 183	51 50 245	22 20 270	38 38 323
251	48 00 065	37 29 116	58 38 156	21 31 184	51 10 246	21 35 271	38 12 323
252	48 40 065	38 09 116	58 55 158	21 27 185	50 29 247	20 51 271	37 45 323
253	49 20 065	38 49 117	59 11 160	21 23 186	49 48 248	20 06 272	37 18 324
254	50 01 066	39 28 118	59 26 162	21 19 187	49 06 249	19 21 273	36 52 324
	♦DENEB	ALTAIR	Nunki	♦ANTARES	ARCTURUS	♦Alkaid	Kochab
255	50 42 066	40 07 119	16 47 153	21 13 188	48 24 250	56 04 299	54 33 346
256	51 22 066	40 46 120	17 06 154	21 07 189	47 42 251	55 25 300	54 21 345
257	52 03 067	41 24 121	17 25 155	21 00 190	47 00 252	54 47 300	54 10 345
258	52 44 067	42 02 122	17 44 156	20 52 190	46 18 253	54 08 300	53 58 345
259	53 25 067	42 40 123	18 01 157	20 43 191	45 35 254	53 29 300	53 46 344
260	54 06 067	43 17 124	18 18 158	20 34 192	44 52 254	52 51 300	53 34 344
261	54 48 068	43 53 125	18 35 159	20 24 193	44 09 255	52 12 300	53 21 344
262	55 29 068	44 29 127	18 51 160	20 14 194	43 26 256	51 34 301	53 09 343
263	56 10 068	45 05 128	19 06 161	20 02 195	42 43 257	50 56 301	52 56 343
264	56 52 068	45 40 129	19 20 162	19 50 196	41 59 258	50 17 301	52 43 343
265	57 33 069	46 14 130	19 34 162	19 38 197	41 15 259	49 39 301	52 29 343
266	58 15 069	46 48 131	19 47 163	19 24 198	40 32 259	49 01 301	52 16 342
267	58 56 069	47 21 132	20 00 164	19 10 199	39 48 260	48 23 302	52 02 342
268	59 38 069	47 54 134	20 11 165	18 55 200	39 04 261	47 45 302	51 48 342
269	60 20 070	48 26 135	20 22 166	18 40 201	38 20 262	47 07 302	51 34 342

LHA ♈	Hc Zn	Hc Zn	Hc Zn	Hc Zn	Hc Zn	Hc Zn	Hc Zn
	♦Alpheratz	ALTAIR	Nunki	♦ANTARES	ARCTURUS	♦Alkaid	Kochab
270	17 40 067	48 57 136	20 33 167	18 24 202	37 36 262	46 30 302	51 20 341
271	18 21 067	49 28 137	20 42 168	18 07 203	36 51 263	45 52 303	51 06 341
272	19 02 068	49 58 139	20 51 169	17 50 203	36 07 264	45 15 303	50 51 341
273	19 43 068	50 27 140	20 59 170	17 32 204	35 23 265	44 37 303	50 37 341
274	20 25 069	50 55 141	21 07 171	17 13 205	34 38 265	44 00 303	50 22 341
275	21 06 069	51 22 143	21 14 172	16 54 206	33 54 266	43 23 304	50 07 340
276	21 48 070	51 49 144	21 20 173	16 34 207	33 09 267	42 46 304	49 52 340
277	22 30 071	52 15 146	21 25 174	16 13 208	32 25 267	42 09 304	49 37 340
278	23 12 071	52 39 147	21 29 175	15 52 209	31 40 268	41 32 305	49 22 340
279	23 54 072	53 03 149	21 33 176	15 31 209	30 56 269	40 55 305	49 06 340
280	24 37 072	53 26 150	21 36 177	15 08 210	30 11 269	40 19 305	48 51 340
281	25 19 073	53 47 152	21 38 178	14 46 211	29 27 270	39 43 305	48 35 340
282	26 02 073	54 08 153	21 40 179	14 22 212	28 42 271	39 06 306	48 20 339
283	26 45 074	54 27 155	21 41 180	13 58 213	27 57 271	38 30 306	48 04 339
284	27 28 075	54 46 157	21 41 180	13 34 214	27 13 272	37 54 306	47 48 339
	♦Mirfak	Alpheratz	♦ALTAIR	Rasalhague	♦ARCTURUS	Alkaid	Kochab
285	13 21 033	28 11 075	55 03 158	55 08 219	26 28 273	37 18 307	47 33 339
286	13 45 033	28 54 076	55 19 160	54 40 220	25 44 273	36 43 307	47 17 339
287	14 10 034	29 37 076	55 34 162	54 11 222	24 59 274	36 07 307	47 01 339
288	14 35 034	30 21 077	55 47 163	53 41 223	24 15 275	35 32 308	46 45 339
289	15 00 035	31 04 077	55 59 165	53 10 224	23 30 275	34 57 308	46 28 339
290	15 25 035	31 48 078	56 10 167	52 38 226	22 46 276	34 22 308	46 12 339
291	15 51 036	32 31 079	56 20 168	52 06 227	22 02 277	33 47 309	45 56 339
292	16 17 036	33 15 079	56 28 170	51 33 228	21 17 277	33 12 309	45 40 339
293	16 44 037	33 59 080	56 35 172	50 59 230	20 33 278	32 38 310	45 24 339
294	17 10 037	34 43 080	56 41 174	50 25 231	19 49 279	32 04 310	45 08 339
295	17 38 038	35 27 081	56 45 176	49 50 232	19 05 279	31 30 310	44 51 339
296	18 05 038	36 11 081	56 47 177	49 14 233	18 21 280	30 56 311	44 35 339
297	18 33 039	36 55 082	56 49 179	48 38 235	17 37 280	30 22 311	44 19 339
298	19 00 039	37 39 083	56 49 181	48 01 236	16 53 281	29 48 311	44 02 339
299	19 29 040	38 23 083	56 47 183	47 24 237	16 10 282	29 15 312	43 46 339
	Mirfak	♦Alpheratz	Enif	♦ALTAIR	Rasalhague	Alphecca	♦Kochab
300	19 57 040	39 08 084	50 38 137	56 44 185	46 47 238	34 27 277	43 30 339
301	20 26 040	39 52 084	51 07 139	56 40 186	46 09 239	33 42 277	43 13 339
302	20 55 041	40 36 085	51 36 140	56 34 188	45 30 240	32 58 278	42 57 339
303	21 24 041	41 21 086	52 04 142	56 27 190	44 51 241	32 14 279	42 41 339
304	21 54 042	42 05 086	52 32 143	56 19 192	44 12 242	31 30 279	42 25 339
305	22 24 042	42 50 087	52 58 144	56 09 193	43 33 243	30 46 280	42 09 339
306	22 54 043	43 34 088	53 24 146	55 58 195	42 53 244	30 02 280	41 52 339
307	23 24 043	44 19 088	53 48 147	55 46 197	42 12 245	29 18 281	41 36 339
308	23 55 044	45 03 089	54 11 149	55 32 199	41 32 246	28 35 282	41 20 339
309	24 26 044	45 48 089	54 34 151	55 17 200	40 51 247	27 51 282	41 04 339
310	24 57 044	46 33 090	54 55 152	55 01 202	40 10 248	27 07 283	40 48 339
311	25 28 045	47 17 091	55 16 154	54 44 204	39 28 249	26 24 283	40 32 339
312	26 00 045	48 02 091	55 35 155	54 26 205	38 47 250	25 41 284	40 16 339
313	26 31 046	48 46 092	55 53 157	54 06 207	38 05 251	24 57 284	40 00 339
314	27 03 046	49 31 093	56 09 159	53 45 208	37 23 251	24 14 285	39 45 339
	CAPELLA	♦Alpheratz	FOMALHAUT	♦ALTAIR	Rasalhague	VEGA	♦Kochab
315	11 11 036	50 15 094	13 24 154	53 23 210	36 40 252	62 38 275	39 29 339
316	11 37 037	51 00 094	13 43 155	53 01 211	35 58 253	61 54 276	39 13 340
317	12 04 037	51 44 095	14 02 156	52 37 213	35 15 254	61 10 276	38 58 340
318	12 31 038	52 29 096	14 20 157	52 12 214	34 32 255	60 25 277	38 42 340
319	12 58 038	53 13 097	14 37 158	51 46 216	33 49 255	59 41 277	38 27 340
320	13 26 039	53 57 097	14 54 158	51 20 217	33 06 256	58 57 278	38 11 340
321	13 54 039	54 41 098	15 10 159	50 52 219	32 22 257	58 13 278	37 56 340
322	14 23 040	55 25 099	15 25 160	50 24 220	31 39 258	57 29 279	37 41 340
323	14 51 040	56 09 100	15 40 161	49 55 221	30 55 259	56 45 279	37 26 340
324	15 20 041	56 53 101	15 54 162	49 25 223	30 11 259	56 01 280	37 11 340
325	15 50 041	57 37 102	16 08 163	48 54 224	29 27 260	55 17 280	36 56 341
326	16 19 042	58 21 103	16 21 164	48 23 225	28 43 261	54 33 281	36 41 341
327	16 49 042	59 04 104	16 33 165	47 51 227	27 59 262	53 49 281	36 27 341
328	17 19 043	59 47 105	16 45 165	47 18 228	27 15 262	53 06 282	36 12 341
329	17 50 043	60 30 106	16 56 166	46 45 229	26 31 263	52 22 282	35 58 341
	♦CAPELLA	Hamal	Diphda	♦FOMALHAUT	ALTAIR	♦VEGA	Kochab
330	18 21 044	36 13 090	19 09 139	17 06 167	46 11 230	51 38 283	35 43 341
331	18 52 044	36 57 091	19 38 140	17 15 168	45 36 231	50 55 283	35 29 342
332	19 23 045	37 42 091	20 07 141	17 24 169	45 01 232	50 12 284	35 15 342
333	19 55 045	38 27 092	20 35 142	17 32 170	44 26 234	49 28 284	35 01 342
334	20 26 046	39 11 093	21 02 143	17 40 171	43 50 235	48 45 285	34 47 342
335	20 58 046	39 56 093	21 29 143	17 47 172	43 13 236	48 02 285	34 34 342
336	21 31 047	40 40 094	21 55 144	17 53 173	42 36 237	47 19 285	34 20 342
337	22 03 047	41 25 095	22 21 145	17 58 173	41 58 238	46 36 286	34 07 343
338	22 36 048	42 09 095	22 46 146	18 03 174	41 21 239	45 53 286	33 54 343
339	23 10 048	42 53 096	23 10 147	18 07 175	40 42 240	45 10 287	33 40 343
340	23 43 049	43 38 097	23 34 148	18 10 176	40 03 241	44 28 287	33 28 343
341	24 16 049	44 22 098	23 58 149	18 13 177	39 24 242	43 45 288	33 15 343
342	24 50 049	45 06 098	24 20 150	18 15 178	38 45 243	43 03 288	33 02 344
343	25 24 050	45 50 099	24 42 151	18 16 179	38 05 244	42 21 289	32 50 344
344	25 58 050	46 34 100	25 03 152	18 16 180	37 25 245	41 38 289	32 37 344
	♦CAPELLA	ALDEBARAN	Diphda	♦FOMALHAUT	ALTAIR	♦VEGA	Kochab
345	26 33 051	15 33 082	25 24 153	18 16 181	36 45 245	40 56 290	32 25 344
346	27 07 051	16 17 082	25 44 154	18 15 182	36 04 246	40 14 290	32 13 344
347	27 42 052	17 01 083	26 03 155	18 13 183	35 23 247	39 32 290	32 01 345
348	28 17 052	17 45 084	26 22 156	18 11 184	34 42 248	38 51 291	31 49 345
349	28 53 053	18 30 084	26 40 157	18 08 184	34 00 249	38 09 291	31 38 345
350	29 28 053	19 14 085	26 57 158	18 04 185	33 18 250	37 28 292	31 26 345
351	30 04 053	19 59 085	27 13 159	18 00 186	32 36 251	36 46 292	31 15 346
352	30 40 054	20 43 086	27 29 160	17 54 187	31 54 251	36 05 293	31 04 346
353	31 16 054	21 27 087	27 44 161	17 48 188	31 12 252	35 24 293	30 53 346
354	31 52 055	22 12 087	27 58 162	17 42 189	30 29 253	34 43 294	30 43 346
355	32 28 055	22 57 088	28 11 163	17 35 190	29 46 254	34 02 294	30 32 347
356	33 05 055	23 41 089	28 24 164	17 27 191	29 04 255	33 22 295	30 22 347
357	33 42 056	24 26 089	28 36 165	17 18 192	28 20 255	32 41 295	30 12 347
358	34 19 056	25 10 090	28 47 166	17 08 193	27 37 256	32 01 296	30 02 347
359	34 56 057	25 55 091	28 57 167	16 58 193	26 54 257	31 21 296	29 52 348

TABLE 5—CORRECTION FOR PRECESSION AND NUTATION

L.H.A. ♈	North latitudes							0°	South latitudes							L.H.A. ♈
	N. 89°	N. 80°	N. 70°	N. 60°	N. 50°	N. 40°	N. 20°		S. 20°	S. 40°	S. 50°	S. 60°	S. 70°	S. 80°	S. 89°	
1982																
0	1 010	1 030	1 040	1 050	1 060	1 060	2 070	2 070	2 070	1 060	1 060	1 050	1 040	1 020	1 000	0
30	1 040	1 050	1 060	1 060	2 070	2 070	2 070	2 070	2 070	1 060	1 050	1 040	1 020	1 350	1 330	30
60	1 060	1 070	1 080	1 080	2 080	2 080	2 080	2 080	1 080	1 070	1 060	0 —	0 —	0 —	1 300	60
90	1 090	1 090	1 090	1 090	2 090	2 090	2 090	2 090	1 090	1 090	1 100	0 —	0 —	0 —	1 270	90
120	1 120	1 110	1 110	1 110	2 110	2 100	2 100	2 100	1 110	1 110	1 120	1 140	0 —	1 220	1 240	120
150	1 150	1 140	1 130	1 120	1 120	2 110	2 110	2 110	2 110	1 120	1 130	1 140	1 160	1 180	1 210	150
180	1 180	1 160	1 140	1 130	1 130	1 120	2 120	2 110	2 120	1 120	1 120	1 130	1 140	1 160	1 170	180
210	1 210	1 190	1 160	1 140	1 130	1 120	2 110	2 110	2 110	2 110	2 110	1 120	1 120	1 130	1 150	210
240	1 240	0 —	0 —	0 —	1 120	1 110	1 100	2 100	2 100	2 100	2 100	1 100	1 110	1 110	1 120	240
270	1 270	0 —	0 —	0 —	1 090	1 090	1 090	2 090	2 090	2 090	2 090	1 090	1 090	1 090	1 090	270
300	1 310	1 320	0 —	1 040	1 060	1 070	1 070	2 080	2 080	2 080	2 080	1 070	1 070	1 070	1 060	300
330	1 340	1 000	1 020	1 040	1 050	1 060	2 070	2 070	2 070	2 070	1 060	1 060	1 050	1 040	1 030	330
360	1 010	1 030	1 040	1 050	1 060	1 060	2 070	2 070	2 070	1 060	1 060	1 050	1 040	1 020	1 000	360
1983																
0	1 000	1 020	1 040	2 050	2 060	2 060	3 070	3 070	3 070	2 060	2 060	2 050	1 040	1 020	1 000	0
30	1 030	1 050	2 060	2 060	2 070	2 070	3 070	3 070	2 070	2 060	1 050	1 040	1 020	1 350	1 330	30
60	1 060	1 070	2 070	2 080	2 080	3 080	3 080	3 080	2 080	1 070	1 060	1 040	1 000	1 320	1 300	60
90	1 090	1 090	2 090	2 090	2 090	3 090	3 090	2 090	2 090	1 090	1 090	0 —	0 —	1 270	1 270	90
120	1 120	1 110	2 110	2 100	2 100	3 100	3 100	2 100	2 110	2 110	1 120	1 140	1 180	1 220	1 240	120
150	1 150	1 130	2 120	2 120	2 120	2 110	3 110	3 110	2 110	2 120	1 130	1 140	1 160	1 190	1 210	150
180	1 180	1 160	1 140	2 130	2 120	2 120	3 120	3 110	3 120	2 120	2 120	2 130	1 140	1 160	1 180	180
210	1 210	1 190	1 160	1 140	1 130	2 120	2 110	3 110	3 110	2 110	2 120	2 120	2 130	1 140	1 150	210
240	1 240	1 220	1 180	1 140	1 120	1 110	◄2 110	3 100	3 100	3 100	2 100	2 110	2 110	1 110	1 120	240
270	1 270	1 270	0 —	0 —	1 090	1 090	2 090	2 090	3 090	3 090	2 090	2 090	2 090	1 090	1 090	270
300	1 300	1 320	1 000	1 040	1 060	1 070	2 080	2 080	3 080	3 080	2 080	2 080	2 070	1 070	1 060	300
330	1 330	1 350	1 020	1 040	1 050	2 060	2 070	3 070	3 070	2 070	2 070	2 060	2 060	1 050	1 030	330
360	1 000	1 020	1 040	2 050	2 060	2 060	3 070	3 070	3 070	2 060	2 060	2 050	1 040	1 020	1 000	360
1984																
0	1 000	1 020	2 040	2 050	2 060	3 060	3 070	4 070	3 070	3 060	3 060	2 050	2 040	2 020	1 010	0
30	1 030	2 040	2 050	3 060	3 060	3 070	3 070	4 070	3 070	2 060	2 050	2 040	1 020	1 000	1 330	30
60	1 060	2 070	2 070	3 070	3 080	3 080	4 080	3 080	3 070	2 070	1 060	1 040	1 000	1 320	1 300	60
90	1 090	2 090	2 090	3 090	3 090	3 090	4 090	3 090	3 090	2 090	1 090	0 —	0 —	1 270	1 270	90
120	1 120	2 110	2 110	3 100	3 100	3 100	4 100	3 100	3 100	2 110	1 120	1 140	1 190	1 230	1 240	120
150	1 150	2 130	2 120	3 120	3 110	3 110	4 110	3 110	3 110	2 120	2 130	2 140	1 160	1 190	1 210	150
180	1 180	2 160	2 140	2 130	3 120	3 120	3 120	4 110	3 120	3 120	2 130	2 130	2 140	1 160	1 180	180
210	1 210	1 180	1 160	2 140	2 130	2 120	3 110	3 110	3 110	3 110	3 120	3 120	2 130	2 140	1 150	210
240	1 240	1 220	1 180	1 140	1 120	2 110	3 110	3 100	4 100	3 100	3 100	3 110	2 110	2 110	1 120	240
270	1 270	1 270	0 —	0 —	1 090	2 090	3 090	3 090	4 090	3 090	3 090	3 090	2 090	2 090	1 090	270
300	1 300	1 320	1 000	1 040	1 060	2 070	3 080	3 080	4 080	3 080	3 080	3 080	2 070	2 070	1 060	300
330	1 330	1 350	1 020	2 040	2 050	2 060	3 070	3 070	4 070	3 070	3 070	3 060	2 060	2 050	1 030	330
360	1 000	1 020	2 040	2 050	2 060	3 060	3 070	4 070	3 070	3 060	3 060	2 050	2 040	2 020	1 010	360
1985																
0	2 000	2 020	2 040	3 050	3 060	4 060	4 070	4 070	4 070	4 060	3 060	3 050	2 040	2 020	2 010	0
30	2 030	2 040	3 050	3 060	4 060	4 070	4 070	4 070	4 070	3 060	3 050	2 040	2 020	2 000	2 340	30
60	2 060	2 070	3 070	3 070	4 080	4 080	4 080	4 080	3 070	2 070	2 060	1 040	1 000	1 320	2 300	60
90	2 090	2 090	3 090	4 090	4 090	4 090	4 090	4 090	3 090	2 090	1 090	1 080	0 —	1 280	2 270	90
120	2 120	2 110	3 110	3 100	4 100	4 100	4 100	4 100	3 100	2 110	2 120	1 140	1 190	1 230	2 240	120
150	2 150	2 130	3 120	3 120	4 110	4 110	4 110	4 110	4 110	3 120	2 130	2 140	2 160	1 190	2 210	150
180	2 180	2 160	2 140	3 130	3 120	4 120	4 120	4 110	4 120	4 120	3 130	3 130	2 140	2 160	2 180	180
210	2 210	2 180	2 160	2 140	3 130	3 120	4 110	4 110	4 110	4 110	4 120	3 120	3 130	2 140	2 150	210
240	2 240	1 220	1 180	1 140	2 120	2 110	3 110	4 100	4 100	4 100	4 100	3 110	3 110	2 110	2 120	240
270	2 270	1 270	0 —	1 100	1 090	2 090	3 090	4 090	4 090	4 090	4 090	4 090	3 090	2 090	2 090	270
300	2 300	1 310	1 350	1 040	2 060	2 070	3 080	4 080	4 080	4 080	4 080	3 080	3 070	2 070	2 060	300
330	2 330	1 350	2 020	2 040	2 050	3 060	4 070	4 070	4 070	4 070	4 070	3 060	3 060	2 050	2 040	330
360	2 000	2 020	2 040	3 050	3 060	4 060	4 070	4 070	4 070	4 060	3 060	3 050	2 040	2 020	2 010	360

Example. In 1984 a position line is obtained in latitude S.52° when L.H.A.♈ is 327°. Entering the table with the year 1984, latitude S.50°, and L.H.A.♈ 330° gives 3′ 070° which indicates that the position line is to be transferred 3 miles in true bearing 070°.

FORM FOR REDUCTION BY H.O. 249

Navigator ....

Observation   Sun ..a.. Planet ....
Moon..... Star ....

TO FIND GMT

Zone Watch Time.............{may be

Z.D. ....{omitted

GMT Watch Time ..*10..51.17* s

GMT Watch Error............*50*   Sub if fast / Add if slow

      GMT   *10. 52. 07*

Greenwich Date .*August 12* 19*83*

N.A.TO FIND LHA (For Sun, Planets, Moon)

p.159 GHA *10* hrs. *328 43.7'* v .....

p. XXVIII Corr. *52m 07s* ...*13 01.8* (yellow

v Corr^n .... table)

(For Stars)

SHA Star ....(daily

GHA ϓ ...hrs................. page)

Corr...m... s ....(yellow

→ GHA .*341° 45.5'*   table)

Assumed Long. ...*29 45.5* Sub if W / Add if E

     LHA   *312° 00.0'*

Zone Date *August 12* 19*83*

D.R. Lat. *42° 53'* (N)(or) S

D.R. Long. *29 40'* E (or) (W)

REDUCTION OF SEXTANT OBSERVATION

Moon's H.P. _____

	+	−
Ht. of Eye *77* ft		
Index Corr.	*2.5*	......
Ht. of Eye Corr.	......	*8.5* (cover
3rd. Corr. to Hs	*14.9*	...... table)
Additional Corr.		
Sums	*17.4*	*8.5* '

Hs .*40° 30.9'*

Total Corr.   + ...*8.9* '   Add if + / Sub. if −

→ Hₒ   *40° 39.8'*

TO FIND DECLINATION

Dec. (daily page) *15° 05.2' N* d *0.7*

d Corr^n − ....*0.6* (yellow table)

DECLINATION   *15° 04.6' N*

Dec. Difference ......*4.6*..

Dec. same name ☑ or contrary name ☐ to Lat.

note sign of d in H.O. 249

SOLUTION BY H.O. 249

(Assumed Position Method).

     LHA *312°* }

Assumed Lat. *N 43°* } H.O. 249

Hₒ .*40° 40'*

H_c .*40 33*

Hₒ− H_c + ...*07'* (If Hₒ is the greater, Hₒ− H_c is + ; + is toward)

→ Tabulated H(Alt.) .*40° 29'*... +d.*42* z *109°*

d (Dec. Diff.) Corr. + ...*4*

      H_c   *40° 33'*

Z_n *109°*

DIRECTIONS:

1. Choose assumed long., close to D.R. Long., so that L.H.A. becomes a whole number of degrees.

2. Choose assumed lat. equal to nearest whole degree to D.R. Lat.

3. N.Hem. For L.H.A. between 0° and 180°: $Z_n = 360° - Z$
For L.H.A. between 180° and 360°: $Z_n = Z$

     S.Hem. For L.H.A. between 0° and 180°: $Z_n = 180° + Z$
For L.H.A. between 180° and 360°: $Z_n = 180° - Z$

4. (a) Plot assumed lat. and long.

(b) Lay down bearing of observed body (Z_n).

(c) From assumed position plot intercept, Hₒ-H_c, toward or away from bearing of body, according to whether Hₒ-H_c is + or -.

(d) Through point found under Step (c) draw a line of position at right angles to the bearing Z_n.

FORM 2

FORM FOR REDUCTION OF NOON SIGHT FROM NOON CURVE

TO FIND GMT OF NOON
from NOON CURVE

GMT Watch Time $13^h 56^m 57^s$

GMT Watch Error .... 50 Sub. if fast / Add if slow

GMT ...13...57...47

Greenwich Date August 12 1983

N.A. DEC. AT NOON

p. 159 Dec. (daily page) 15 02.9'   d 0.7

p. XXX d Corr. & sign 0.7

DEC. AT NOON 15° 02.2' N

DERIVATION OF NOON LATITUDE

RULE: Find zenith distance of sun by subtracting H_O from 90°. Mark it N if zenith is N of sun, mark it S if zenith is S of sun. If zenith distance and declination are both N or both S add them, subtract if one is N, other S. The result is the latitude with the name of the greater.

$$90° = 89° 60.0'$$

Repeat H_O = 62° 20.0 Sub.

Zenith Distance = 27° 40.0 (N) or S

Repeat Dec. at Noon = 15 02.2 (N) or S

LATITUDE AT NOON = 42° 42.2' (N) or S

Zone Date August 12 19 83

D.R. Lat. 42° 41.1' (N) (or) S

D.R. Long. 28° 0.8' E (or) (W)

True Course 100° Speed 22 knots.

REDUCTION OF SEXTANT OBSERVATION

	+	−	
Ht. of Eye 7.7 ft.			
Index Corr.	0.5		
Ht. of Eye Corr.		8.5	(cover
3rd. Corr to h_s	15.5		table)
Additional Corr.			
Sums	16.0	8.5	

H_s 62° 12.5'

Total Corr. 7.5 Add if + / Sub. if −

→ H_O 62° 20.0'

DERIVATION OF NOON LONGITUDE

FIND GHA

GHA 13 hrs. 13° 44.0' (daily page)

Corr. 57m 47s 14 26.8 (yellow

GHA 28° 10.8' table)

= LONG. W if less than 180°

If greater than 180°, 360° − GHA

= LONG. E   = _____ E

FORM 3

FORM FOR REDUCTION OF NOON SIGHT   Navigator ....

TO FIND GMT	Zone Date ...Augus.t...12.... 19.83.
Zone Watch Time .... {may be	D.R. Lat. 42.°.51' N (or) S
Z. D. .... {omitted	D.R. Long. 29.°24.' E (or) W
GMT Watch Time 11.ʰ.24.ᵐ.00.ˢ	True Course 100.° Speed 22. knots.
GMT Watch Error ....50 Sub. if fast / Add if slow	D Lo/hr. 30. E (or) W (Mins. of Arc)
GMT 11.24..50.	**REDUCTION OF SEXTANT OBSERVATION**
Greenwich Date August 12. 19.83	

PREDICTED WATCH TIME OF NOON

N.A. **TO FIND LHA** ° '	Ht. of Eye 77. ft.
p. 159 GHA 11. hrs. ...343.°43.8 (daily page)	
p. XiV Corr. 24ᵐ50ˢ ......6.°12.5 (yellow	
GHA 349.°56.3' table)	
D.R. Long. 29.°24.0 Sub. if W / Add if E	
LHA 320.°32.3'	

Reduction of sextant observation table:

Ht. of Eye 77. ft.	+	−	
Index Corr.	0.5		
Ht. of Eye Corr.		8.5	(cover
3rd. Corr to Hₛ	15.5		table)
Additional Corr.	......		
Sums	16.0	8.5	
Hₛ	62°12.5'		
Total Corr. +	7.5		Add if + / Sub. if −
H₀	62°20.0'		

PREDICTED WATCH TIME OF NOON

360° = 359° 60.0'

> when vessel's speed is greater than 12 knots

LHA = 320° 32.3' Sub.	
t = 39° 27.7' E	
Corr. from Table at bottom of Form = 1° 16.4' Add if + / Sub. if −	
38° 11.3'	
Interval to Noon 2ʰ 32ᵐ 45ˢ ↑Arc to ↓Time	
Repeat Watch Time 11.24.00 Add	
PRED. W.T. of NOON 13ʰ56ᵐ45ˢ	

DERIVATION OF NOON LATITUDE

RULE: Find zenith distance of sun by subtracting H₀ from 90°. Mark if N if zenith is N of sun, mark it S if zenith is S of sun. If zenith distance and declination are both N or both S add them, subtract if one is N, other S. The result is the latitude with the name of the greater.

PREDICTED GMT & DEC. AT NOON

Repeat Interval to Noon 2ʰ32ᵐ45ˢ	
Repeat GMT & Add 11.24.50	
Predicted GMT of Noon 13ʰ57ᵐ35ˢ	
Dec.(daily page) 15°02.9' N d − 7	
d Corrⁿ & sign 0.7	
DEC. AT NOON 15°02.2' N ↗	

90° = 89° 60.0'	
Repeat H₀ = 62°20.0 Sub.	
Zenith Distance = 27°40.0 N (or) S	
Repeat Dec. at Noon = 15°02.2 N (or) S	
LATITUDE AT NOON = 42°42.2' N (or) S	

OMIT ↓   **TABLE OF CORR. TO t**

It is generally unnecessary to apply a correction when vessel's speed is less than 12 knots. At greater speeds rough interpolation in the table and rounding off is permissible. For practical purposes an error of 10' in the value of the correction will lead to an error of only 40ˢ in the predicted time of noon.

DLo/hr. / t	60' E	50' E	40' E	30' E	20' E	15' E	10' E	5' E	0	5' W	10' W	15' W	20' W	30' W	40' W	50' W	60' W
10°	− 37'.5	− 31'.6	− 25'.5	− 19'.4	− 13'.0	− 9'.8	− 6'.6	− 3'.3	0	+ 3'.4	+ 6'.8	+10'.2	+13'.6	+ 20'.7	+ 27'.9	+ 35'.3	+ 42'.9
20°	− 75'.0	− 63'.2	− 51'.1	− 38'.7	− 26'.1	− 19'.7	− 13'.2	− 6'.6	0	+ 6'.7	+13'.5	+20'.3	+27'.3	+ 41'.4	+ 55'.8	+ 70'.6	+ 85'.7
30°	− 112'.5	− 94'.7	− 76'.6	− 58'.1	− 39'.1	− 29'.5	− 19'.8	− 9'.9	0	+10'.1	+20'.2	+30'.5	+40'.9	+ 62'.1	+ 83'.7	+105'.9	+128'.6
40°	− 150'.0	− 126'.3	− 102'.1	− 77'.4	− 52'.2	− 39'.3	− 26'.4	− 13'.3	0	+13'.4	+27'.0	+40'.7	+54'.5	+ 82'.8	+111'.6	+141'.2	+171'.4
50°	− 187'.5	− 157'.9	− 127'.7	− 96'.8	− 65'.2	− 49'.2	− 33'.0	− 16'.6	0	+16'.8	+33'.7	+50'.8	+68'.2	+103'.5	+139'.5	+176'.5	+214'.3
60°	− 225'.0	− 189'.5	− 153'.2	− 116'.1	− 78'.3	− 59'.0	− 39'.6	− 19'.9	0	+20'.1	+40'.4	+61'.0	+81'.8	+124'.1	+167'.4	+211'.9	+257.1

FORM 4
VOL. II & VOL. III

FORM FOR REDUCTION BY H.O. 249

Navigator ....
Observation Sun .2. Planet ....
Moon ..... Star ....

TO FIND GMT

Zone Watch Time.............. { may be
Z. D. .... { omitted

Zone Date ..August. 12.. 19.83.

D.R. Lat. 42° 34' (N) (or) S

D.R. Long. 26° 58' E (or) (W)

GMT Watch Time .16.h 20.m 25.s

GMT Watch Error.............51. Sub.if fast / Add if slow

GMT ..16. 21. ..16.

Greenwich Date August 12, 1983.

REDUCTION OF SEXTANT OBSERVATION

Moon's H.P. _____

	+	−
Ht. of Eye .7.7. ft		
Index Corr.	......	0..5.
Ht. of Eye Corr.		8..5. (cover
3rd. Corr. to Hs	15..1	...... table)
Additional Corr.	......	......
Sums	15.1	9.0

N.A. TO FIND LHA (For Sun, Planets, Moon)

p.159 GHA 16 hrs. ..58. 44.3' 𝑣 .....

p. Xii Corr 21.m 16.s ..5. 19.0. (yellow table)

𝑣 Corr" .... table)

(For Stars)

SHA Star ....(daily

GHA Υ ...hrs................. page)

Corr...m... s ....(yellow table)

GHA ....64° 03.3'

Assumed Long. .27. 03.3 Sub.if W / Add if E

LHA .37° 00.0'

Hs ..47. 53.0'

Total Corr. ....6.1 Add if + / Sub. if −

Ho 47° 59.1'

TO FIND DECLINATION

Dec. (daily page) 15° 00.7' N d 0.7

d Corr" − ....0.3. (yellow table)

DECLINATION 15° 00.4' N

SOLUTION BY H.O. 249
(Assumed Position Method)

LHA 37° }
Assumed Lat. 42 N } H.O. 249

Ho 47° 59
Hc 48 17
Ho − Hc − 18'

Dec. Difference .....0.0....
Dec. same name ☑ or contrary name ☐ to Lat.

note sign of d in H.O. 249

Tabulated H(Alt.) ....48° 17' + .44 z 119°
d (Dec. Diff.) Corr. ....0........

Hc 48° 17' Zn 241°

(If Ho is the greater, Ho − Hc is + ; + is toward)

DIRECTIONS:

1. Choose assumed long., close to D.R. Long., so that L.H.A. becomes a whole number of degrees.

2. Choose assumed lat. equal to nearest whole degree to D.R. Lat.

3. N.Hem. For L.H.A. between 0° and 180°: $Z_n = 360° - Z$
   For L.H.A. between 180° and 360°: $Z_n = Z$

   S.Hem. For L.H.A. between 0° and 180° : $Z_n = 180° + Z$
   For L.H.A. between 180° and 360°: $Z_n = 180° - Z$

4. (a) Plot assumed lat. and long.

   (b) Lay down bearing of observed body (Z_n).

   (c) From assumed position plot intercept, $H_0 - H_c$, toward or away from bearing of body, according to whether $H_0 - H_c$ is + or −.

   (d) Through point found under Step (c) draw a line of position at right angles to the bearing Z_n.

FORM 5

FORM FOR REDUCTION BY H.O. 249

VOL. II a VOL. III

Navigator ....

Observation: Sun ...... Planet ....
Moon. �
⃝. Star ....

TO FIND GMT

Zone Watch Time .... ⎰may be
Z. D. ....⎱omitted

GMT Watch Time .. $16^h \ 21^m \ 55^s$ ⎰Sub if fast

GMT Watch Error .... 51 ⎱Add if slow

GMT .. 16. 22. 46.

Greenwich Date August 12 1983

N.A. TO FIND LHA (For Sun, Planets, Moon)

p.159 GHA ... hrs. ..... 6° 24.9' 𝑣 11.3

p.xiii Corr ...m ...s .. 5. 25. 9. (yellow

p.xiii 𝑣 Corr^n .... 4.2.. table)

(For Stars)

SHA Star ....(daily

GHA ♈ ...hrs .... page)

Corr ...m ... s ....(yellow
table)

GHA .11° 55. 0'

Assumed Long. 2.6. 5.5. 0 ⎰Sub if W
⎱Add if E

LHA 345° 00.0'

Zone Date August 12 .... 19.83.

D.R. Lat. 42° 34' Ⓝ(or) S

D.R. Long. 26° 58' E (or) Ⓦ

REDUCTION OF SEXTANT OBSERVATION

Moon's H.P. 52.3'

Ht. of Eye 7.7 ft

	+	—
Index Corr.	......	1.5
Ht. of Eye Corr.		8.5 (cover
3rd. Corr. to Hs	51.1	table)
Additional Corr.	4.4	30.0
Sums	55.5	40.0

H_s .. 44° 13.6'

Total Corr. + ... 15.5 ⎰Add if +
⎱Sub. if —

H_O .. 44° 29.1'

TO FIND DECLINATION

Dec. (daily page) .. 0° 50.2' S d 14.7

d Corr^n + ... 5.5 (yellow table)

DECLINATION .. 0° 55.7' S

SOLUTION BY H.O. 249
(Assumed Position Method)

LHA 345°
Assumed Lat. N 42° ⎱ H.O. 249

H_O .. 44° 29'

H_C .. 44 59' ⟵

$H_O - H_C$ .... 30'

Dec. Difference ... 5.6

Dec. same name ☐ or contrary name ☐ to Lat.

note sign of d in H.O. 249

⟶ Tabulated H (Alt.) .. 45° 53' ⁻ d 5.8 Z 15.8°

d (Dec. Diff.) Corr.⁻ 5.4

H_C .. 44° 59' Z_n 158°

(If H_O is the greater, $H_O - H_C$ is + ; + is toward)

DIRECTIONS:

1. Choose assumed long., close to D.R. Long., so that L.H.A. becomes a whole number of degrees.

2. Choose assumed lat. equal to nearest whole degree to D.R. Lat.

3. N.Hem. For L.H.A. between 0° and 180°: Z_n = 360° - Z
   For L.H.A. between 180° and 360°: Z_n = Z

   S.Hem. For L.H.A. between 0° and 180° : Z_n = 180° + Z
   For L.H.A. between 180° and 360°: Z_n = 180° - Z

4. (a) Plot assumed lat. and long.

   (b) Lay down bearing of observed body (Z_n).

   (c) From assumed position plot intercept, $H_O - H_C$, toward or away from bearing of body, according to whether $H_O - H_C$ is + or -.

   (d) Through point found under Step (c) draw a line of position at right angles to the bearing Z_n.

FORM 6
STAR FIX (H.O. 249)
(Vol. 1)

WATCH TIME 21.14

D.R. Lat 42.13' (N) (or) S Zone Date

D.R. Long 24°39' E (or) (W) Aug. 12

| STAR _Arcturus_ | STAR _Antares_ | STAR _Altair_ |

REDUCTION OF SEXTANT OBSERVATION

	+	−			+	−			+	−	
Ht. of Eye 7.7 ft											
Index Corr.	2.5	......			2.5	......			2.5	......	
Ht. of Eye Corr.		8.5	(cover			8.5	(cover			8.5	(cover
3rd. Corr. to Hs		0.9	table)			2.5	table)			1.1	table)
Sums	2.5	9.4			2.5	11.0			2.5	9.6	

Hs 48° 45.3'   Total Corr. − 6.9 (Add if +, Sub if −)   Ho 48° 38.4'

Hs 21° 07.0'   − 8.5   Ho 20° 58.5'

Hs 40° 39.0'   − 7.1   Ho 40° 31.9'

TO FIND GMT

Zone Watch Time .... {may be omitted	Zone Watch Time	Zone Watch Time	
Z.D.	Z.D.	Z.D.	
GMT Watch Time 21ʰ 12ᵐ 50ˢ	**GMT** Watch Time 21ʰ 16ᵐ 08ˢ	**GMT** Watch Time 21ʰ 17ᵐ 25ˢ	
GMT Watch Error 53 (Sub if fast, Add if slow)	GMT Watch Error 53	GMT Watch Error 53	
GMT 21 13 43	**GMT** 21 17 01	**GMT** 21 18 18	
Greenwich Date Aug. 12 1983	Greenwich Date Aug. 12	Greenwich Date Aug. 12	
GHA ♈ 21 hrs 275° 44.4' (daily page)	GHA ♈ 21 hrs 275° 44.4'	GHA ♈ 21 hrs 275° 44.4'	
Corr 13.m 43 s 3 26.3 (yellow table)	Corr 17.m 01 s 4 15.9	Corr 18.m 18 s 4 35.3	
GHA ♈ 279° 10.7'	GHA ♈ 280° 00.3'	GHA ♈ 280° 19.7'	
Assumed Long 24 10.7 (Sub if W, Add if E)	Assumed Long 25 00.3	Assumed Long 24 19.7	
LHA ♈ 255° 00.0'	LHA ♈ 255° 00.0'	LHA ♈ 256° 00.0'	

Assumed Lat. 42° N	Ass. Lat. 42° N	Ass. Lat. 42° N
Repeat Ho 48° 38.4' (from above)	Repeat Ho 20° 58.5 (from above)	Repeat Ho 40° 31.9 (from above)
Hc (H.O.249) 48 24	Hc (H.O.249) 21 13	Hc (H.O.249) 40 46
Ho − Hc + 14'	Ho − Hc − 15'	Ho − Hc − 14'
Zn (H.O.249) 250.0°	Zn (H.O.249) 188°	Zn (H.O.249) 120°

DIRECTIONS FOR PLOTTING:

(a) Plot assumed lat. and long.

(b) Lay down bearing of observed body (Zn).

(c) From assumed position plot intercept, H₀-Hc, toward or away from bearing of body, according to whether H₀-Hc is + or −.

(If H₀ is the greater, H₀-Hc is +).

(d) Through point found under Step (c) draw a line of position at right angles to the bearing Zn.

(e) Apply PRECESSION & NUTATION CORR, if necessary, to FIX or to L.O.P.'s.

FORM 7
STAR FIX (H.O. 249)
(Vol. 1)

WATCH TIME .21.17...

D.R. Lat..42°.13'. (N)(or) S Zone Date

D.R. Long.24°.38'. E (or) (W). Aug. 12.

STAR *Arcturus*   STAR *Antares*   STAR *Altair*

REDUCTION OF SEXTANT OBSERVATION

Ht. of Eye .7.7. ft

	+	−	
Index Corr.	2.5	......	
Ht. of Eye Corr.		8.5	(cover
3rd. Corr. to H_s		0.9	table)
Sums	2.5	9.4	

H_s 48° 51.7'

Total Corr. ...... 6.9   Add if +
                        Sub. if −

H_o 48° 44.8'

	+	−	
	2.5	......	
		8.5	(cover
		2.5	table)
	2.5	11.0	

H_s 21° 07.0'

− 8.5

H_o 20° 58.5'

	+	−	
	2.5	......	
		8.5	(cover
		1.1	table)
	2.5	9.6	

H_s 41° 05.7'

− 7.1

H_o 40° 58.6'

TO FIND GMT

Zone Watch Time .... { may be
Z.D. .... { omitted

GMT Watch Time ....
GMT Watch Error........... Sub if fast
                          Add if slow

GMT .21^h.13^m.01^s

Greenwich Date *Aug. 12*.19....

GHA ♈ ...hrs..............(daily page)
Corr...m... s .... (yellow table)
GHA ♈ ....
Assumed Long. .... Sub if W
                            Add if E
LHA ♈ 254° 00.0'

Assumed Lat. .42°.N...
Repeat H_o 48° 45'
(from above)
H_c (H.O.249) 49 06

$H_o − H_c$ − 21'

Z_n (H.O.249)..... 249°

TO FIND GMT

Zone Watch Time.............
Z.D. ....
GMT Watch Time 21^h.16^m.08
GMT Watch Error........... 53

GMT 21^h.17^m.01^s

Greenwich Date *Aug. 12*

GHA ♈ 21 hrs 275° 44.4'
Corr 17.m 01 s 4 15.9
GHA ♈ 280° 00.3'
Assumed Long. 25 00.3
LHA ♈ 255° 00.0'

Ass. Lat. 42° N
Repeat H_o 20° 58'
(from above)
H_c (H.O.249) 21 13

$H_o − H_c$ − 15'

Z_n (H.O.249) 188°

TO FIND GMT

Zone Watch Time ....
Z.D. ....
GMT Watch Time ....
GMT Watch Error...........

GMT 21^h 21^m 01^s

Greenwich Date *Aug. 12*

GHA ♈ ...hrs..............
Corr...m... s ....
GHA ♈ ....
Assumed Long. ....
LHA ♈ 256° 00.0'

Ass. Lat. 42° N
Repeat H_o 40° 59'
(from above)   40 46
H_c (H.O.249)

$H_o − H_c$ + 13'

Z_n (H.O.249) 120°

DIRECTIONS FOR PLOTTING:

(a) Plot assumed lat. and long.

(b) Lay down bearing of observed body (Z_n).

(c) From assumed position plot intercept, $H_0 − H_c$, toward or away from bearing of body, according to whether $H_0 − H_c$ is + or −.

(If H_0 is the greater, $H_0 − H_c$ is +).

(d) Through point found under Step (c) draw a line of position at right angles to the bearing Z_n.

(e) Apply PRECESSION & NUTATION CORR, if necessary, to FIX or to L.O.P.'s.

FORM 8　　　　　VOL. II & VOL. III

FORM FOR REDUCTION BY H.O. 249

Navigator ....
Observation { Sun...... Planet *Saturn* ... Moon..... Star }

TO FIND GMT

Zone Watch Time............... } may be
Z. D. .... } omitted

GMT Watch Time 21ʰ 12ᵐ 57ˢ
GMT Watch Error .... 53 ⎰ Sub if fast ⎱ Add if slow

GMT 21 13 50

Greenwich Date *August 12* 19 83

N.A. **TO FIND LHA** (For Sun, Planets, Moon)
p. 158 GHA 21 hrs. 67° 47.4 𝒗 2.3
p.Viii Corr. 13m 50s 3 27.5 (yellow
p.Viii 𝒗 Corrⁿ .... 5 table)

(For Stars)

SHA Star .... (daily
GHA ϒ ... hrs. .... page)
Corr... m ... s .... (yellow
GHA 71° 15.4 table)
Assumed Long. 24 15.4 ⎰ Sub. if W ⎱ Add if E
LHA 47° 00.0'

Zone Date *August 12* 19 83
D.R. Lat. 42° 13' (N) (or) S
D.R. Long. 24° 39' E (or) (W)

REDUCTION OF SEXTANT OBSERVATION

	+	−	Moon's H.P. ____
Ht. of Eye 7.7 ft			
Index Corr.	2.5	......	
Ht. of Eye Corr.	......	8.5	(cover
3rd. Corr. to Hs	......	2.2	table)
Additional Corr.	......	......	
Sums	2.5	10.7	

Hs 23° 34.3'
Total Corr. − .... 8.2 ⎰ Add if + ⎱ Sub. if −
Ho 23° 26.1'

TO FIND DECLINATION

Dec. (daily page) 8° 57.8' S d 0.1
d Corrⁿ .... 0.0 (yellow table)
DECLINATION 8° 57.8' S

SOLUTION BY H.O. 249
(Assumed Position Method)

LHA 47°
Assumed Lat. N 42° } H.O. 249

Dec. Difference 58
Dec. same name ☐ or contrary name ☑ to Lat.
note sign of d in H.O. 249

↳ Tabulated H (Alt.) 24° 08' d 42 z 128°
d (Dec. Diff.) Corr. − 47

H₀ 23° 26'
H_c 23 21' ←——— H_c 23° 2.1' z_n 232°
H₀−H_c + ... 5 (If H₀ is the greater, H₀−H_c is + ; + is toward)

DIRECTIONS:

1. Choose assumed long., close to D.R. Long., so that L.H.A. becomes a whole number of degrees.

2. Choose assumed lat. equal to nearest whole degree to D.R. Lat.

3. N.Hem. For L.H.A. between 0° and 180°: Z_n = 360° − Z
   For L.H.A. between 180° and 360°: Z_n = Z

   S.Hem. For L.H.A. between 0° and 180°: Z_n = 180° + Z
   For L.H.A. between 180° and 360°: Z_n = 180° − Z

4. (a) Plot assumed lat. and long.
   (b) Lay down bearing of observed body (Z_n).
   (c) From assumed position plot intercept, H₀−H_c, toward or away from bearing of body, according to whether H₀−H_c is + or −.
   (d) Through point found under Step (c) draw a line of position at right angles to the bearing Z_n.

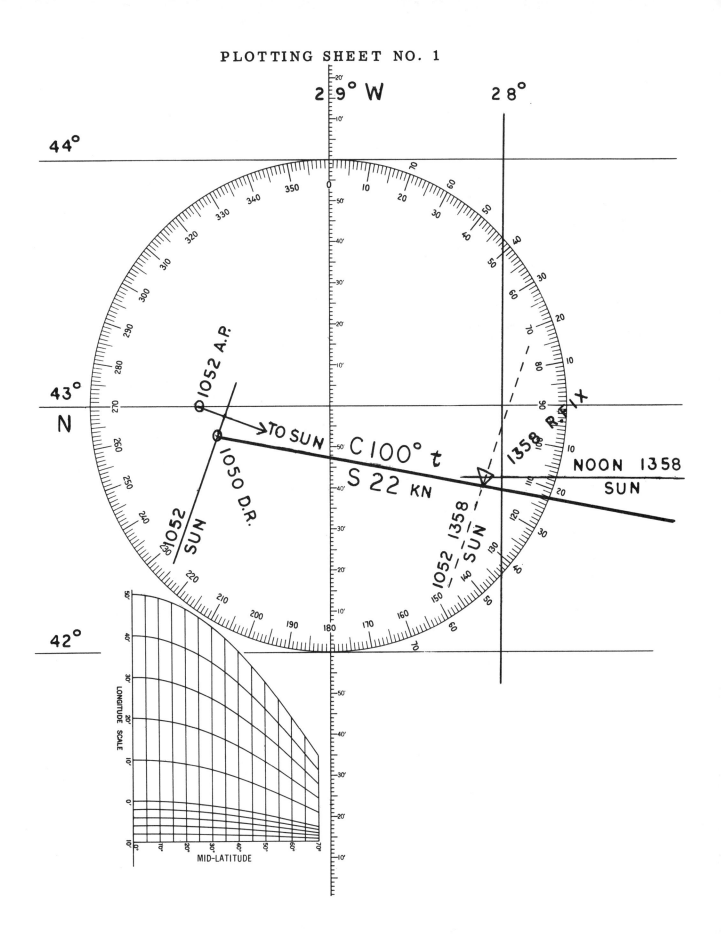

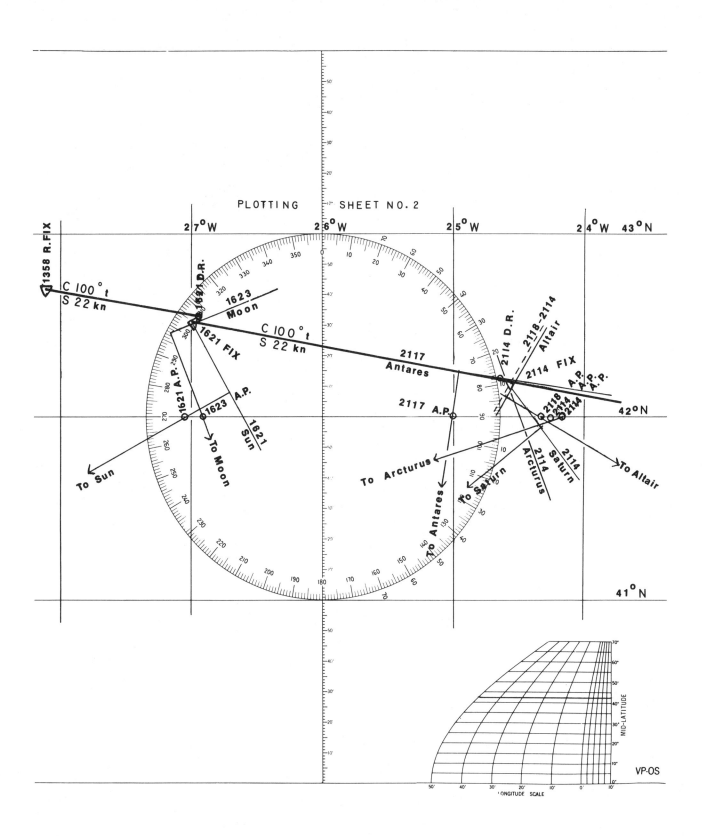

PLOTTING SHEET NO. 2

FORM FOR REDUCTION BY H.O. 249

Navigator ....

Observation { Sun ...... Planet ....
Moon..... Star ....

TO FIND GMT

Zone Watch Time............. { may be
Z. D. .... { omitted

GMT Watch Time .... Sub if fast

GMT Watch Error.................. Add if slow

GMT ....

Greenwich Date .... 19....

TO FIND LHA (For Sun, Planets, Moon)

GHA ...hrs. .... v ....

Corr...^{m} ...^{s} ....(yellow

v Corr^n .... table)

(For Stars)

SHA Star ....(daily

GHA ♈ ...hrs................. page)

Corr...^{m}... ^{s} ....(yellow table)

GHA ....

Assumed Long. .... { Sub. if W
Add if E

LHA ....$000'$

Zone Date .... 19......

D.R. Lat. .... N (or) S

D.R. Long. .... E (or) W

REDUCTION OF SEXTANT OBSERVATION

Moon's H.P. _____

	+	−	
Ht. of Eye ..... ft.			
Index Corr.	......	......	
Ht. of Eye Corr.		......	(cover
3rd. Corr. to H_s	......		table)
Additional Corr.	......	......	
Sums			

H_s ....

Total Corr. .... { Add if +
Sub. if −

H_o ....

TO FIND DECLINATION

Dec. (daily page)................ d ....

d Corr^n ....(yellow table)

DECLINATION ....

SOLUTION BY H.O. 249

(Assumed Position Method)

LHA ....
Assumed Lat. .... } H.O. 249

H_o ....

H_c .... ⟵ H_c ....

$H_o - H_c$ .... (If H_o is the greater, $H_o - H_c$ is + ; + is toward)

Dec. Difference ....

Dec. same name ☐ or contrary name ☐ to Lat.

note sign of d in H.O. 249

Tabulated H (Alt.) .... d..... Z ....

d (Dec. Diff.) Corr. ....

Z_n ....

DIRECTIONS :

1. Choose assumed long., close to D.R. Long., so that L.H.A. becomes a whole number of degrees.

2. Choose assumed lat. equal to nearest whole degree to D.R. Lat.

3. N. Hem. For L.H.A. between 0° and 180°: $Z_n = 360° - Z$
   For L.H.A. between 180° and 360°: $Z_n = Z$

   S. Hem. For L.H.A. between 0° and 180° : $Z_n = 180° + Z$
   For L.H.A. between 180° and 360°: $Z_n = 180° - Z$

4. (a) Plot assumed lat. and long.

   (b) Lay down bearing of observed body (Z_n).

   (c) From assumed position plot intercept, $H_o - H_c$, toward or away from bearing of body, according to whether $H_o - H_c$ is + or -.

   (d) Through point found under Step (c) draw a line of position at right angles to the bearing Z_n.

FORM FOR REDUCTION OF NOON SIGHT FROM NOON CURVE

TO FIND GMT OF NOON
from NOON CURVE

GMT Watch Time ....

GMT Watch Error .... Sub. if fast
_____ Add if slow

GMT ....

Greenwich Date .... 19....

DEC. AT NOON

Dec. (daily page) .... d ...

d Corr. & sign ....

DEC. AT NOON ....

DERIVATION OF NOON LATITUDE

RULE: Find zenith distance of sun by subtracting H_O from 90°. Mark it N if zenith is N of sun, mark it S if zenith is S of sun. If zenith distance and declination are both N or both S add them, subtract if one is N, other S. The result is the latitude with the name of the greater.

$$90° = 89° \ 60.0'$$

Repeat H_O = ... Sub. ←

Zenith Distance = ... N (or) S

Repeat Dec. at Noon = ... N (or) S

LATITUDE AT NOON = ... N (or) S

Zone Date .... 19.......

D.R. Lat. .... N (or) S

D.R. Long. .... E (or) W

True Course .... Speed ..... knots.

REDUCTION OF SEXTANT OBSERVATION

	+	−	
Ht. of Eye .... ft.			
Index Corr.	......	......	
Ht. of Eye Corr.		......	(cover
3rd. Corr to h_s	......		table)
Additional Corr.	......		
Sums			

H_s .... Add if +

Total Corr. .... Sub. if −

→ H_O ....

DERIVATION OF NOON LONGITUDE

FIND GHA

GHA ... hrs. .... (daily page)

Corr. ...^{m} ...^{s} .... (yellow

GHA .... table)

= LONG. W if less than 180°

If greater than 180°, 360° − GHA

= LONG. E = _____ E

106

FORM FOR REDUCTION OF NOON SIGHT Navigator ....

TO FIND GMT

Zone Watch Time .... ⎰may be

Z. D. .... ⎱omitted

GMT Watch Time ....

GMT Watch Error .... Sub.if fast
————————————— Add if slow

GMT ....

Greenwich Date .... 19....

TO FIND LHA

GHA ... hrs. .... (daily page)

Corr. ...ᵐ ...ˢ .... (yellow
——————————————— table)

GHA ....

D.R. Long. .... Sub. if W
——————————————— Add if E

LHA ....

PREDICTED WATCH TIME OF NOON

360° = 359° 60.0'

> when vessel's speed
> is greater than 12 knots

LHA = ... Sub.

t = ... E

Corr. from Table .... Add if +
at bottom of Form ————————————— Sub. if −

.... ↑ Arc to
——————————————— ↓ Time

Interval to Noon ....

Repeat Watch Time .... Add

PRED. W.T. of NOON ....

PREDICTED GMT & DEC. AT NOON

Repeat Interval to Noon ....

Repeat GMT & Add ....

Predicted GMT of Noon ....

Dec. (daily page) .... d ...

d Corrⁿ & sign ....

DEC. AT NOON ....

Zone Date .... 19......

D.R. Lat. .... N (or) S

D.R. Long. .... E (or) W

True Course .... Speed ..... knots.

D Lo/hr. .... E (or) W (Mins. of Arc)

REDUCTION OF SEXTANT OBSERVATION

	+	−	
Ht. of Eye .... ft.			
Index Corr.	......	......	
Ht. of Eye Corr.		......	(cover
3rd. Corr to Hₛ	......		table)
Additional Corr.	......		
Sums			

Hₛ .... Add if +

Total Corr. .... Sub. if −

Hₒ ....

DERIVATION OF NOON LATITUDE

RULE: Find zenith distance of sun by subtracting Hₒ from 90°. Mark it N if zenith is N of sun, mark it S if zenith is S of sun. If zenith distance and declination are both N or both S add them, subtract if one is N, other S. The result is the latitude with the name of the greater.

90° = 89° 60.0'

Repeat Hₒ = ... Sub.

Zenith Distance = ... N (or) S

Repeat Dec. at Noon = ... N (or) S

LATITUDE AT NOON = ... N (or) S

OMIT ↘ TABLE OF CORR. TO t

It is generally unnecessary to apply a correction when vessel's speed is less than 12 knots. At greater speeds rough interpolation in the table and rounding off is permissible. For practical purposes an error of 10' in the value of the correction will lead to an error of only 40ˢ in the predicted time of noon.

DLo/hr. t	60' E	50' E	40' E	30' E	20' E	15' E	10' E	5' E	0	5' W	10' W	15' W	20' W	30' W	40' W	50' W	60' W
10°	− 37'.5	− 31'.6	− 25'.5	− 19'.4	− 13'.0	− 9'.8	− 6'.6	− 3'.3	0	+ 3'.4	+ 6'.8	+10'.2	+13'.6	+ 20'.7	+ 27'.9	+ 35'.3	+ 42'.9
20°	− 75'.0	− 63'.2	− 51'.1	− 38'.7	− 26'.1	− 19'.7	− 13'.2	− 6'.6	0	+ 6'.7	+13'.5	+20'.3	+27'.3	+ 41'.4	+ 55'.8	+ 70'.6	+ 85'.7
30°	−112'.5	− 94'.7	− 76'.6	− 58'.1	− 39'.1	− 29'.5	− 19'.8	− 9'.9	0	+10'.1	+20'.2	+30'.5	+40'.9	+ 62'.1	+ 83'.7	+105'.9	+128'.6
40°	−150'.0	−126'.3	−102'.1	− 77'.4	− 52'.2	− 39'.3	− 26'.4	− 13'.3	0	+13'.4	+27'.0	+40'.7	+54'.5	+ 82'.8	+111'.6	+141'.2	+171'.4
50°	−187'.5	−157'.9	−127'.7	− 96'.8	− 65'.2	− 49'.2	− 33'.0	− 16'.6	0	+16'.8	+33'.7	+50'.8	+68'.2	+103'.5	+139'.5	+176'.5	+214'.3
60°	−225'.0	−189'.5	−153'.2	−116'.1	− 78'.3	− 59'.0	− 39'.6	− 19'.9	0	+20'.1	+40'.4	+61'.0	+81'.8	+124'.1	+167'.4	+211'.3	+257.1

FORM FOR REDUCTION BY H.O. 249

Navigator ....

Observation
Sun...... Planet ....
Moon..... Star ....

TO FIND GMT

Zone Watch Time.............. {may be
Z. D. ....{omitted

GMT <u>Watch Time</u> ....

GMT Watch Error.................... Sub if fast / Add if slow

GMT ....

Greenwich Date .... 19....

Zone Date .... 19.....

D.R. Lat. .... N (or) S

D.R. Long. .... E (or) W

REDUCTION OF SEXTANT OBSERVATION

TO FIND LHA (For Sun, Planets, Moon)

GHA ... hrs. .... v .....

Corr...^{m} ...^{s} ....(yellow

v Corr^n .... table)

(For Stars)

SHA Star ....(daily

GHA Υ ...hrs................. page)

Corr...^{m}... ^{s} ....(yellow table)

GHA .... Sub. if W

Assumed Long. .... Add if E

LHA ....$\underset{\cdot\cdot}{0}\underset{\cdot\cdot}{0}\underset{\cdot\cdot}{0}.'$

	+	–	Moon's H.P. ____
Ht. of Eye ..... ft			
Index Corr.	......	......	
Ht. of Eye Corr.		......	(cover
3rd. Corr. to H$_s$	......	......	table)
Additional Corr.	......	......	
Sums			

H$_s$ ....

Total Corr. .... Add if + Sub. if –

$\rightarrow$ H$_0$ ....

TO FIND DECLINATION

Dec. (daily page)................ d ....

d Corr^n ....(yellow table)

DECLINATION ....

SOLUTION BY H.O. 249

(Assumed Position Method)

LHA ....
Assumed Lat. ....} H.O. 249

H$_0$ ....

H$_c$ ....

H$_0$– H$_c$ .... (If H$_0$ is the greater, H$_0$– H$_c$ is + ; + is toward)

Dec. Difference ....

Dec. same name ☐ or contrary name ☐ to Lat.

note sign of d in H.O. 249

$\rightarrow$ Tabulated H(Alt.) .... d..... Z ....

d (Dec. Diff.) Corr. ....

H$_c$ .... Z$_n$ ....

DIRECTIONS :

1. Choose assumed long., close to D.R. Long., so that L.H.A. becomes a whole number of degrees.

2. Choose assumed lat. equal to nearest whole degree to D.R. Lat.

3. N.Hem. For L.H.A. between 0° and 180°: Z$_n$ = 360° - Z
   For L.H.A. between 180° and 360°: Z$_n$ = Z

   S.Hem. For L.H.A. between 0° and 180° : Z$_n$ = 180° + Z
   For L.H.A. between 180° and 360°: Z$_n$ = 180° - Z

4. (a) Plot assumed lat. and long.

   (b) Lay down bearing of observed body (Z$_n$).

   (c) From assumed position plot intercept, H$_0$-H$_c$, toward or away from bearing of body, according to whether H$_0$-H$_c$ is + or -.

   (d) Through point found under Step (c) draw a line of position at right angles to the bearing Z$_n$.

FORM FOR REDUCTION BY H.O. 249

Navigator ....

Observation
Sun...... Planet ....
Moon..... Star ....

TO FIND GMT

Zone Watch Time.............. } may be
Z. D. ....} omitted

GMT Watch Time ....
GMT Watch Error.................. Sub if fast
Add if slow

GMT ....

Greenwich Date .... 19....

TO FIND LHA (For Sun, Planets, Moon)

GHA ...hrs. .... v .....
Corr...m ...s .... (yellow
v Corr^n .... table)

(For Stars)

SHA Star .... (daily
GHA ♈ ...hrs .... page)
Corr...m... s .... (yellow
GHA .... table)
Sub. if W
Assumed Long. .... Add if E
LHA ....0.0.0.'

Zone Date .... 19......
D.R. Lat. .... N (or) S
D.R. Long. .... E (or) W

REDUCTION OF SEXTANT OBSERVATION

Moon's H.P. _____

	+	−	
Ht. of Eye ..... ft			
Index Corr.	......	......	
Ht. of Eye Corr.		......	(cover
3rd. Corr. to Hs	......	......	table)
Additional Corr.	......	......	
Sums			

Hs ....
Total Corr. .... Add if +
Sub. if −
Ho ....

TO FIND DECLINATION

Dec. (daily page)................ d ....
d Corr^n .... (yellow table)
DECLINATION ....

SOLUTION BY H.O. 249

(Assumed Position Method)

LHA ....
Assumed Lat. .... } H.O. 249

Dec. Difference ....
Dec. same name □ or contrary name □ to Lat.

note sign of d in H.O. 249

Tabulated H (Alt.) .... d..... Z ....
d (Dec. Diff.) Corr. ....

Ho ....
Hc _____ ←———— Hc .... Zn ....

Ho − Hc .... (If Ho is the greater, Ho − Hc is + ; + is toward)

DIRECTIONS :

1. Choose assumed long., close to D.R. Long., so that L.H.A. becomes a whole number of degrees.

2. Choose assumed lat. equal to nearest whole degree to D.R. Lat.

3. N. Hem. For L.H.A. between 0° and 180°: $Z_n = 360° - Z$
For L.H.A. between 180° and 360°: $Z_n = Z$

   S. Hem. For L.H.A. between 0° and 180° : $Z_n = 180° + Z$
For L.H.A. between 180° and 360°: $Z_n = 180° - Z$

4. (a) Plot assumed lat. and long.

   (b) Lay down bearing of observed body (Z_n).

   (c) From assumed position plot intercept, $H_0 - H_c$, toward or away from bearing of body, according to whether $H_0 - H_c$ is + or -.

   (d) Through point found under Step (c) draw a line of position at right angles to the bearing Z_n.

STAR FIX (H.O.249)
(Vol. I)

D.R. Lat.............. N (or) S Zone Date
D.R. Long.............E (or) W ....

STAR _____   STAR _____   STAR _____

REDUCTION OF SEXTANT OBSERVATION

Ht. of Eye ..... ft.

	+	−	
Index Corr.	......	......	
Ht. of Eye Corr.		......	(cover
3rd. Corr. to H_s		......	table)
Sums			

H_S ....

Total Corr. .... Add if + Sub. if −

H_o ....

	+	−	
	......	......	
		......	(cover
		......	table)

H_S ....
....

H_o ....

	+	−	
	......	......	
		......	(cover
		......	table)

H_S ....
....

H_o ....

TO FIND GMT

Zone Watch Time .... {may be omitted}

Z.D. ....

GMT Watch Time ....

GMT Watch Error .... Sub if fast / Add if slow

GMT ....

Greenwich Date .... 19....

GHA ♈ ...hrs.............(daily page)

Corr...m... s ....(yellow table)

GHA ♈ ....

Assumed Long. .... Sub if W / Add if E

LHA ♈ .... 00.0'

TO FIND GMT

Zone Watch Time ....

Z.D. ....

GMT Watch Time ....

GMT Watch Error ....

GMT ....

Greenwich Date ....

GHA ♈ ...hrs.............

Corr...m... s ....

GHA ♈ ....

Assumed Long. ....

LHA ♈ .... 00.0'

TO FIND GMT

Zone Watch Time ....

Z.D. ....

GMT Watch Time ....

GMT Watch Error ....

GMT ....

Greenwich Date ....

GHA ♈ ...hrs.............

Corr...m... s ....

GHA ♈ ....

Assumed Long. ....

LHA ♈ .... 00.0'

Assumed Lat. ....

Repeat H_o ....
(from above)

H_C (H.O.249) ....

$H_o - H_C$ ....

Z_n (H.O.249)............

Ass. Lat. ....

Repeat H_o ....
(from above)

H_C (H.O.249) ....

$H_o - H_C$ ....

Z_n (H.O.249)............

Ass. Lat. ....

Repeat H_o ....
(from above)

H_C (H.O.249) ....

$H_o - H_C$ ....

Z_n (H.O.249)............

DIRECTIONS FOR PLOTTING:

(a) Plot assumed lat. and long.

(b) Lay down bearing of observed body (Z_n).

(c) From assumed position plot intercept, H_0-H_C, toward or away from bearing of body, according to whether H_0-H_C is + or -.

(If H_0 is the greater, H_0-H_C is +).

(d) Through point found under Step (c) draw a line of position at right angles to the bearing Z_n.

(e) Apply PRECESSION & NUTATION CORR, if necessary, to FIX or to L.O.P.'s.

STAR FIX (H.O. 249)
(Vol. 1)

STAR _____ STAR _____ STAR _____

REDUCTION OF SEXTANT OBSERVATION

Ht. of Eye ..... ft.

	+	−	
Index Corr.	......	......	
Ht. of Eye Corr.		......	(cover table)
3rd. Corr. to H_s		......	
Sums			

H_s .... Add if +

Total Corr. .... Sub. if −

H_o ....

	+	−	
	......	......	
		......	(cover table)
		......	

H_s ....

....

H_o ....

	+	−	
	......	......	
		......	(cover table)
		......	

H_s...................

....

H_o...................

TO FIND GMT

Zone Watch Time .... {may be omitted}

Z.D. ....

GMT Watch Time ....

GMT Watch Error................... Sub if fast Add if slow

GMT ....

Greenwich Date ....19....

GHA ♈ ...hrs...................(daily page)

Corr...m... s .... (yellow table)

GHA ♈ ....

Assumed Long. .... Sub if W Add if E

LHA ♈ ....0.0.0.'..

TO FIND GMT

Zone Watch Time..............

Z.D. ....

GMT Watch Time ....

GMT Watch Error...................

GMT ....

Greenwich Date ....

GHA ♈ ...hrs..............

Corr...m... s ....

GHA ♈ ....

Assumed Long. ....

LHA ♈ ....0.0.0.'

TO FIND GMT

Zone Watch Time..............

Z.D. ....

GMT Watch Time ....

GMT Watch Error...................

GMT ....

Greenwich Date ....

GHA ♈ ...hrs..............

Corr...m... s ....

GHA ♈ ....

Assumed Long. ....

LHA ♈ ....0.0.0.'

Assumed Lat. ....

Repeat H_o ....
(from above)

H_c (H.O.249) ....

$H_o - H_c$ ....

Z_n (H.O.249)...............

Ass. Lat. ....

Repeat H_o ....
(from above)

H_c (H.O.249) ....

$H_o - H_c$ ....

Z_n (H.O.249)..............

Ass. Lat. ....

Repeat H_o...........
(from above)

H_c (H.O.249)...............

$H_o - H_c$ ....

Z_n (H.O.249)..........

DIRECTIONS FOR PLOTTING:

(a) Plot assumed lat. and long.

(b) Lay down bearing of observed body (Z_n).

(c) From assumed position plot intercept, H_0-H_c, toward or away from bearing of body, according to whether H_0-H_c is + or −.

(If H_0 is the greater, H_0-H_c is +).

(d) Through point found under Step (c) draw a line of position at right angles to the bearing Z_n.

(e) Apply PRECESSION & NUTATION CORR, if necessary, to FIX or to L.O.P.'s.

FORM FOR REDUCTION BY H.O. 249

Navigator ....

Observation Sun...... Planet ....
Moon...... Star..................

TO FIND GMT

Zone Watch Time............ } may be

Z. D. .... omitted

GMT Watch Time ....

GMT Watch Error.................. Sub. if fast
Add if slow

GMT ....

Greenwich Date .... 19....

Zone Date .... 19.....

D.R. Lat. .... N (or) S

D.R. Long. .... E (or) W

REDUCTION OF SEXTANT OBSERVATION

Moon's H.P. _____

	+	−	
Ht. of Eye ..... ft			
Index Corr.	......	......	
Ht. of Eye Corr.		......	(cover
3rd. Corr. to H_s	......	......	table)
Additional Corr.	......	......	
Sums			

TO FIND LHA (For Sun, Planets, Moon)

GHA ... hrs. .... v .....

Corr...^{m} ...^{s} .... (yellow

v Corr^n _____ table)

(For Stars)

SHA Star .... (daily

GHA ♈ ... hrs.............. page)

Corr...^{m} ... ^{s} .... (yellow
table)

GHA ....

Assumed Long. .... Sub. if W
Add if E

LHA .... O O.O'

H_s ....

Total Corr. .... Add if +
Sub. if −

H_o ....

TO FIND DECLINATION

Dec. (daily page)................ d ....

d Corr^n .... (yellow table)

DECLINATION ....

SOLUTION BY H.O. 249

(Assumed Position Method)

LHA .... }
Assumed Lat. .... } H.O. 249

H_o ....

H_c ....

$H_o - H_c$ ....

Dec. Difference ....

Dec. same name ☐ or contrary name ☐ to Lat.

note sign of d in H.O. 249

Tabulated H(Alt.) .... d..... Z ......

d (Dec. Diff.) Corr. ....

H_c .... Z_n ......

(If H_o is the greater, $H_o - H_c$ is +; + is toward)

DIRECTIONS :

1. Choose assumed long., close to D.R. Long., so that L.H.A. becomes a whole number of degrees.

2. Choose assumed lat. equal to nearest whole degree to D.R. Lat.

3. N.Hem. For L.H.A. between 0° and 180°: $Z_n = 360° - Z$
   For L.H.A. between 180° and 360°: $Z_n = Z$

   S.Hem. For L.H.A. between 0° and 180° : $Z_n = 180° + Z$
   For L.H.A. between 180° and 360°: $Z_n = 180° - Z$

4. (a) Plot assumed lat. and long.

   (b) Lay down bearing of observed body (Z_n).

   (c) From assumed position plot intercept, $H_o - H_c$, toward or away from bearing of body, according to whether $H_o - H_c$ is + or -.

   (d) Through point found under Step (c) draw a line of position at right angles to the bearing Z_n.

UNIVERSAL PLOTTING SHEET Navigator

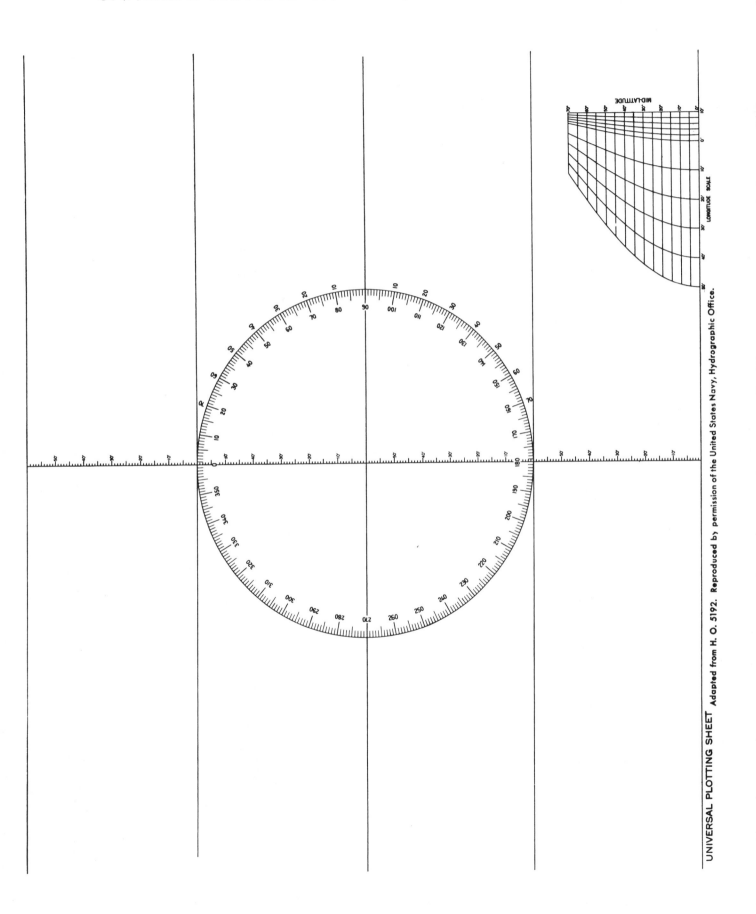

UNIVERSAL PLOTTING SHEET Adapted from H. O. 5192. Reproduced by permission of the United States Navy, Hydrographic Office.

113

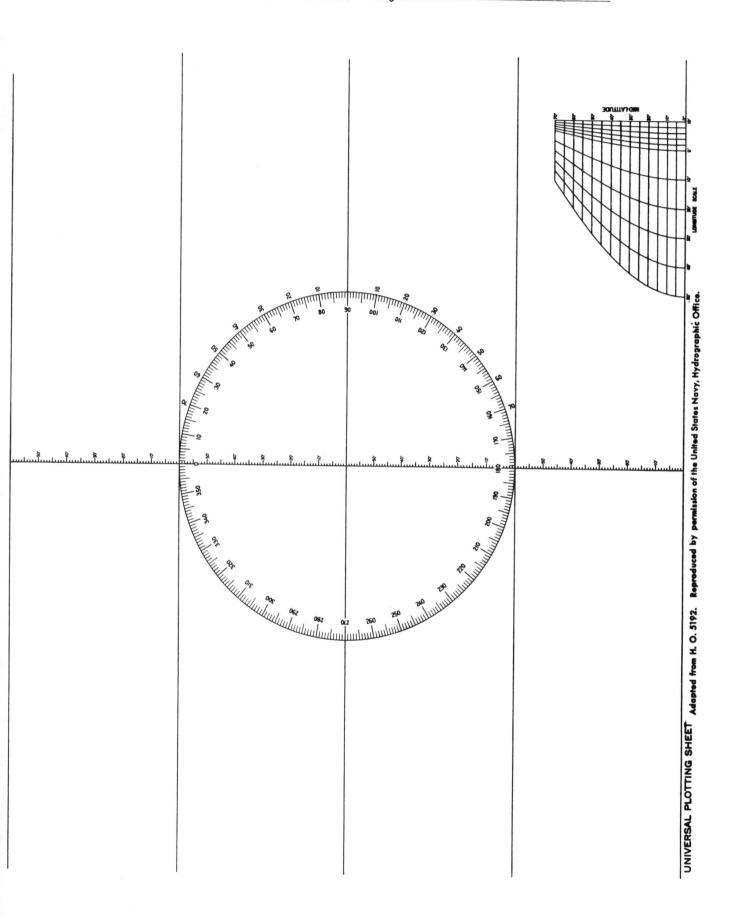

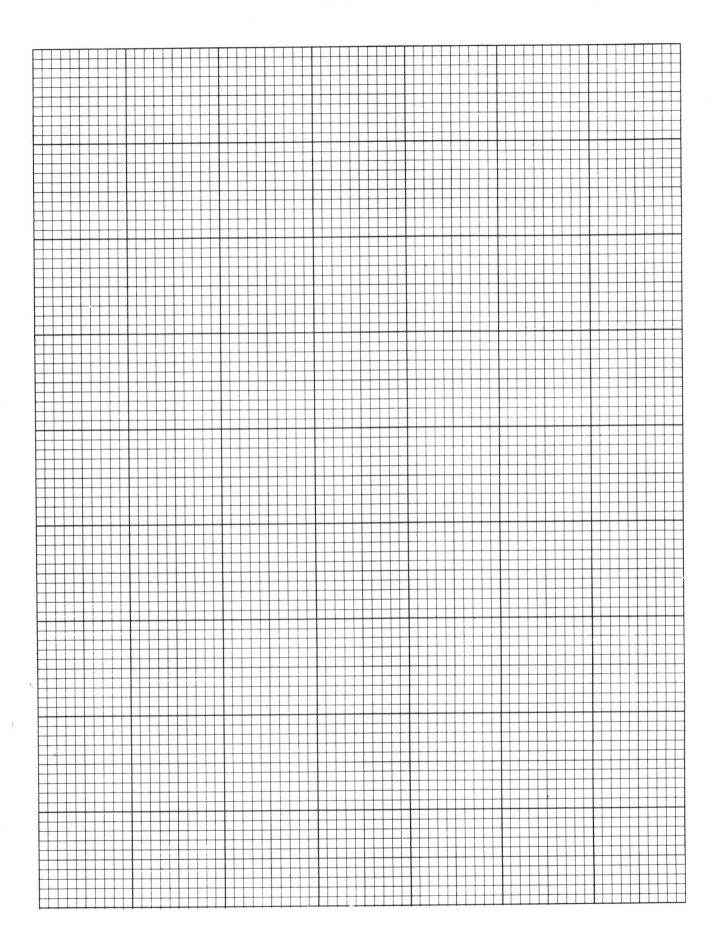

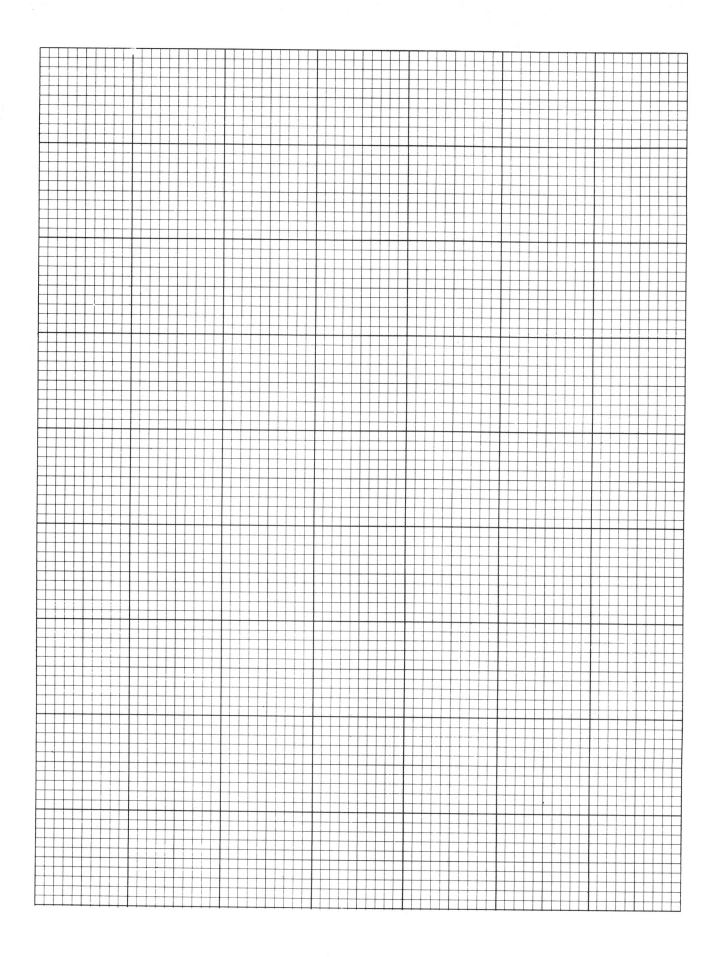

116

PART XI

Abbreviations and Symbols

Ac	Computed altitude	**H. O.**	Hydrographic Office
Alt.	Altitude	**H. P.**	Horizontal parallax
Ao	Measured altitude	**hr.**	Hours of time
App. Alt.	Apparent altitude	**Hs**	Sextant altitude
A. P.	Assumed position	**Ht.**	Height
A2	Altitude corrections	**I. C.**	Index correction
		I. E.	Index error
Cel. equator	Celestial equator		
Cel. meridian	Celestial meridian	**LAN**	Local apparent noon
Chr.	Chronometer	**LAT**	Local apparent time
co Alt.	90° − Altitude	**Lat.**	Latitude
co Lat.	90° − Latitude	**LHA**	Local hour angle
Corr^n	Correction	**LOP**	Line of position
		LMT	Local mean time
d	Declination change in one hour	**Long.**	Longitude
Dec.	Declination		
D Lo	Difference of longitude (in arc units)	**Mer.**	Meridian
		Mer. Pass.	Meridian passage
DMAHC	Defense Mapping Agency Hydrographic Center	**Mid.**	Middle or mean
		N	North
DR	Dead reckoning; Dead reckoning position	**N. A. (on Forms)**	The *Nautical Almanac*
		N cel. pole	North celestial pole
DST	Daylight or summer time		
		Pass.	Passage
E	East	**Pub. No.**	Publication Number
Eqn. of Time	Equation of time		
ETA	Estimated time of arrival	**R Fix**	Running fix
GAT	Greenwich apparent time		
GHA	Greenwich hour angle	**S**	South, on horizon
GMT	Greenwich mean time	**S**	Sun (in Figures)
		S cel. pole	South celestial pole
Hc	Calculated altitude	**ST**	Standard time
Ho	Observed altitude		

117

t	Local hour angle, West or East, never exceeding 180°	**Z**	Zenith; angle in navigation triangle at the zenith
		ZD	Zenith distance
UT	Universal Time	**Z. D.**	Zone description without regard for sign
W	West	**Zn**	Azimuth to be plotted
WT	Watch time or time on the watch	**ZT**	Zone time

♈	First point of Aries; Vernal equinox	**h**	Hours of time
		′	Minutes of arc
β **Cassiopeiae**	Caph	**m**	Minutes of time
α **Crucis**	Acrux	″	Seconds of arc
γ **Crucis**	Gacrux	**s**	Seconds of time
β **Crucis**	Mimosa	>	Greater than
°	Degrees	<	Less than
v	Excess of GHA change for an hour over the adopted value	☋	Lower limb sextant sight
		☊	Upper limb sextant sight

The DUT1 Codes Use Doubled or Emphasized Second Markers to Indicate the Difference Between UT1 and UTC.

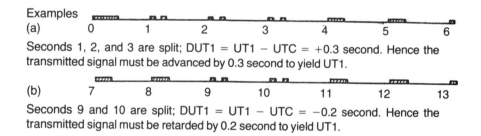

Examples
(a)

Seconds 1, 2, and 3 are split; DUT1 = UT1 − UTC = +0.3 second. Hence the transmitted signal must be advanced by 0.3 second to yield UT1.

(b)

Seconds 9 and 10 are split; DUT1 = UT1 − UTC = −0.2 second. Hence the transmitted signal must be retarded by 0.2 second to yield UT1.

Longitude Correction Due to DUT1	
DUT1	Correction to Longitude
−0ˢ9 to −0ˢ7	0′.2 to east
−0ˢ6 to −0ˢ3	0′.1 to east
−0ˢ2 to +0ˢ2	no correction
+0ˢ3 to +0ˢ6	0′.1 to west
+0ˢ7 to +0ˢ9	0′.2 to west

From "The New UTC Time Signals," by R. L. Duncombe and P. K. Seidelmann. *Navigation* (24:2) 1977.

References

Bayless, Allan, *Compact Sight Reduction Table,* Cornell Maritime Press, 1980

Blythe, J. H., Duncombe, R. L., and Sadler, D. H., "Sight Reduction Tables for Marine Navigation," *Journal of the Institute of Navigation,* 13, No. 2 (Summer 1966).

Bok, Bart J., and Wright, Frances W., *Basic Marine Navigation.* Houghton Mifflin Co., 1952.

Chichester, Francis, *The Lonely Sea and the Sky.* Coward-McCann, Inc., 1964.

Gould, Rupert T., R.N. (Ret.), Lieut. Commander, *John Harrison and His Timekeepers.* National Maritime Museum, 1958.

Hodge, Paul W., *Concepts of Contemporary Astronomy,* McGraw-Hill Book Co., 1979.

Hough, Richard, *Admirals in Collision.* The Viking Press, 1959.

Lansing, Alfred, *Endurance.* McGraw-Hill Book Co., 1959.

Mixter, George W., *Primer of Navigation.* D. Van Nostrand Co., Inc., 1967.

Moscow, Alvin, *Collision Course.* G. P. Putnam's Sons, 1959.

Richey, M. W., "Notes on a Passage to the Azores," *The Journal of the Institute of Navigation,* Vol. 20, No. 1, January 1967, at The Royal Geographical Society, 1 Kensington Gore, London SW7.

Rockefeller, Larry, "Around the World," *Flying,* November 1967.

Seidelmann, P. K. and Feldman, Sidney, "Is there a Future for Celestial Navigaton in the Navy?", *Proceedings of the U. S. Naval Institute,* March 1979.

Shackleton, Ernest, "Boat Journey," from *Great Sea Stories,* selected with introductions by Captain Alan Villiers. Dell Publishing Co., Inc., 1959.

Tabarly, Eric, *Lonely Victory: Atlantic Race 1964.* Clarkson N. Potter, Inc., 1966.

Worsley, F. A., *Shackleton's Boat Journey,* W. W. Norton and Co. Inc., 1977.

Wright, Frances W., *Coastwise Navigation,* Cornell Maritime Press, 1980.

___, *Particularized Navigation, 1973*
Part I, Emergency Booklet
Part II, Emergency Pamphlet

Nautical Almanac for the year. Washington: Government Printing Office. Necessary extracts for Practice Cruise in Part X.

Defense Mapping Agency Hydrographic Center (DMAHC) Publications, Washington, D. C.

American Practical Navigator, H. O. Pub. No. 9 (Bowditch), DMA Pub: NVPUB9V1, V2, 1977, 1975.

Navigation Tables for Mariners and Aviators, H. O. Pub. No. 208 (Dreisonstok), 1942. (out of print)

Dead Reckoning Altitude and Azimuth Table, H. O. Pub. No. 211 (Ageton), 1943. (out of print as a separate volume by U. S. Government; now Table 35 of *Vol. 2 of Pub. 9,* Bowditch or NVPUB9V2. See page 45.)

Tables of Computed Altitude and Azimuth, H. O. Pub. No. 214, 9 vols. (discontinued, but still accurate)

Sight Reduction Tables for Marine Navigation, Pub. 229, 6 vols. (provides the most accurate reduction tables, but is not as fast as *Pub. No. 249,* Vol. 1 for stars. Stock No. SRPUB229V1 for Vol. 1. Each volume covers 15° in latitude.)

Sight Reduction Tables for Air Navigation, Pub. No. 249

Vol. 1 (Selected Stars), 1980, printed every 5 years (Stock No. SRPUB249V1)

Vol. 2 (Latitudes 0°-40°), 1978 (never out of date)

(Declinations 0°-29°) (Stock No. SRPUB249V2)

Vol. 3 (Latitudes 39°-89°), 1967 (never out of date)

(Declinations 0°-29°) (Stock No. SRPUB249V3)

Azimuths of the Sun and Other Celestial Bodies of Declination 0° to 23°, Pub. No. 260, 1964 (never out of date, Stock No. SRPUB260)

Plotting Sheets:

May be purchased from the Defense Mapping Agency Hydrographic Center Distribution Center, 6500 Brookes Lane, Washington, D. C. 20315.

Universal Plotting Sheets:

VP-OS, for use in all latitudes, arranged in pads of fifty (100 charts). Purchasable from Defense Mapping Agency, Office of Distribution Services, 6500 Brookes Lane, Washington, D. C. 20315. (DMA Stock No. VPOSX001)

Position Plotting Sheets, for use in restricted latitudes, as indicated; cost less than for charts with ocean or land features

Latitudes 0°-65° (1) WOBZP (900-910) series (overall size 20 by 38 inches) *Scale: 1° longitude equals 4 inches*

Latitudes 4°S-78° (2) WOBZP (920-936) series (overall size 35 by 46 inches) *Scale: 1° longitude equals 4 inches* (2 inches for 934 and 935)

Latitudes 0°-74° (3) WOXZP (940-953) series (overall size 10 by 18 inches) *Scale: 1° longitude equals 2 inches* (1 inch for 952 and 953) (Designed for use in lifeboats)

Latitudes 2°S-49° (4) WOXZP (960-975) series (overall size 17 by 22 inches) *Scale: 1° longitude equals 4 inches*

A current Defense Mapping Agency *Catalog of Maps, Charts and Related Products* - Part 2 - *Hydrographic Products, Numerical Listing of Nautical Charts and Publications (Pub. 1-N-L)* gives more detailed information about regions or subjects covered by various charts, and also the current prices. This *Catalog* also lists agents (with addresses) for the sale of U. S. ocean and coastal charts and publications, both in the United States and in foreign countries. *It is easiest* to go to a government sales agent for charts and publications. In addition, the agent will be most familiar with the many changes which frequently occur. The mariner has, as his reference to changes, the *Notice to Mariners,* which is a weekly publication by DMAHC. In ordering items from the government, a request must be accompanied by a check or money order, and if an error is made in the price, probably due to a change in costs, the time delay may be doubled. The author strongly advises help from an agent in acquiring charts and publications for a cruise.

Index

FORM FOR REDUCTION OF NOON SIGHT   Navigator ....

TO FIND GMT	

TO FIND GMT

Zone Watch Time .... } may be

Z. D. .... } omitted

GMT Watch Time ....

GMT Watch Error .... Sub. if fast / Add if slow

GMT ....

Greenwich Date .... 19....

TO FIND LHA

GHA ... hrs. ....(daily page)

Corr. ...^m ...^s ....(yellow table)

GHA ....

D.R. Long. .... Sub. if W / Add if E

LHA ....

PREDICTED WATCH TIME OF NOON

$360° = 359° 60.0'$

| when vessel's speed is greater than 12 knots |

LHA = ... Sub.

t = ... E

Corr. from Table at bottom of Form .... Add if + / Sub. if —

.... ↑ Arc to ↓ Time

Interval to Noon ....

Repeat Watch Time .... Add

PRED. W.T. of NOON ....

PREDICTED GMT & DEC. AT NOON

Repeat Interval to Noon ....

Repeat GMT & Add ....

Predicted GMT of Noon ....

Dec. (daily page) .... d ...

d Corr^n & sign ....

DEC. AT NOON ....

Zone Date .... 19......

D.R. Lat. .... N (or) S

D.R. Long. .... E (or) W

True Course .... Speed ..... knots.

D Lo/ hr. .... E (or) W (Mins. of Arc)

REDUCTION OF SEXTANT OBSERVATION

	+	—	
Ht. of Eye .... ft.			
Index Corr.	......	......	
Ht. of Eye Corr.		......	(cover
3rd. Corr to H_s	......		table)
Additional Corr.	......		
Sums			

H_s .... Add if +

Total Corr. .... Sub. if —

H_o ....

DERIVATION OF NOON LATITUDE

RULE: Find zenith distance of sun by subtracting H_o from 90°. Mark it N if zenith is N of sun, mark it S if zenith is S of sun. If zenith distance and declination are both N or both S add them, subtract if one is N, other S. The result is the latitude with the name of the greater.

$90° = 89° 60.0'$

Repeat H_o = ... Sub.

Zenith Distance = ... N (or) S

Repeat Dec. at Noon = ... N (or) S

LATITUDE AT NOON = ... N (or) S

OMIT   TABLE OF CORR. TO t

It is generally unnecessary to apply a correction when vessel's speed is less than 12 knots. At greater speeds rough interpolation in the table and rounding off is permissible. For practical purposes an error of 10' in the value of the correction will lead to an error of only 40s in the predicted time of noon.

DLo/ hr. / t	60' E	50' E	40' E	30' E	20' E	15' E	10' E	5' E	0	5' W	10' W	15' W	20' W	30' W	40' W	50' W	60' W
10°	− 37'.5	− 31'.6	− 25'.5	− 19'.4	− 13'.0	− 9'.8	− 6'.6	− 3'.3	0	+ 3'.4	+ 6'.8	+10'.2	+13'.6	+ 20'.7	+ 27'.9	+ 35'.3	+ 42'.9
20°	− 75'.0	− 63'.2	− 51'.1	− 38'.7	− 26'.1	− 19'.7	− 13'.2	− 6'.6	0	+ 6'.7	+ 13'.5	+20'.3	+27'.3	+ 41'.4	+ 55'.8	+ 70'.6	+ 85'.7
30°	−112'.5	− 94'.7	− 76'.6	− 58'.1	− 39'.1	− 29'.5	− 19'.8	− 9'.9	0	+10'.1	+20'.2	+30'.5	+40'.9	+ 62'.1	+ 83'.7	+105'.9	+128'.6
40°	−150'.0	−126'.3	−102'.1	− 77'.4	− 52'.2	− 39'.3	− 26'.4	− 13'.3	0	+13'.4	+27'.0	+40'.7	+54'.5	+ 82'.8	+ 111'.6	+141'.2	+171'.4
50°	−187'.5	−157'.9	−127'.7	− 96'.8	− 65'.2	− 49'.2	− 33'.0	− 16'.6	0	+16'.8	+33'.7	+50'.8	+68'.2	+103'.5	+139'.5	+176'.5	+214'.3
60°	−225'.0	−189'.5	−153'.2	−116'.1	− 78'.3	− 59'.0	− 39'.6	− 19'.9	0	+20'.1	+40'.4	+61'.0	+81'.8	+124'.1	+167'.4	+211'.9	+257.1

FORM FOR REDUCTION OF NOON SIGHT FROM NOON CURVE

TO FIND GMT OF NOON
from NOON CURVE

GMT Watch Time ....

GMT Watch Error .... Sub. if fast
.... Add if slow

GMT ....

Greenwich Date .... 19....

DEC. AT NOON

Dec.(daily page) .... d ...

d Corr. & sign ....

DEC. AT NOON ....

DERIVATION OF NOON LATITUDE

RULE: Find zenith distance of sun by subtracting H_O from 90°. Mark it N if zenith is N
of sun, mark it S if zenith is S of sun. If
zenith distance and declination are both N or
both S add them, subtract if one is N, other
S. The result is the latitude with the name
of the greater.

$$90° = 89° 60.0'$$

Repeat H_O = ... Sub. ←

Zenith Distance = ... N(or)S

Repeat Dec. at Noon = ... N(or)S

LATITUDE AT NOON = ... N(or)S

Zone Date .... 19.......

D.R. Lat. .... N (or) S

D.R. Long. .... E (or) W

True Course .... Speed ..... knots.

REDUCTION OF SEXTANT OBSERVATION

	+	−	
Ht. of Eye .... ft.			
Index Corr.	......	......	
Ht. of Eye Corr.		......	(cover
3rd. Corr to h_s	......		table)
Additional Corr.	......		
Sums			

H_s .... Add if +

Total Corr. .... Sub. if −

→ H_O ....

DERIVATION OF NOON LONGITUDE

FIND GHA

GHA ... hrs. .... (daily page)

Corr. ...^{m} ...^{s} .... (yellow

GHA .... table)

= LONG. W if less than 180°

If greater than 180°, 360° − GHA

= LONG. E = ‾‾‾‾‾ E

FORM FOR REDUCTION BY H.O. 249

Navigator ....

TO FIND GMT	Observation Sun...... Planet .... Moon..... Star ....
Zone Watch Time............. {may be	Zone Date .... 19......
Z. D. .... {omitted	D.R. Lat. .... N (or) S
GMT Watch Time	D.R. Long. .... E (or) W
GMT Watch Error.................. Sub if fast / Add if slow	**REDUCTION OF SEXTANT OBSERVATION**

REDUCTION OF SEXTANT OBSERVATION   Moon's H.P. _____

	+	−	
Ht. of Eye ..... ft.			
Index Corr.	......	......	
Ht. of Eye Corr.		......	(cover
3rd. Corr. to H_s	......	......	table)
Additional Corr.	......	......	
Sums			

GMT ....
Greenwich Date .... 19....

TO FIND LHA (For Sun, Planets, Moon)
GHA ... hrs. .... v .....
Corr...^{m} ...^{s} ....(yellow
v Corr^n .... table)

(For Stars)
SHA Star ....(daily
GHA ϒ ...hrs.................. page)
Corr...^{m}... ^{s} ....(yellow table)
GHA ....
Assumed Long. .... Sub. if W / Add if E
LHA .......O.O.O.'

H_s .... Add if +
Total Corr. .... Sub. if −
H_o ....

TO FIND DECLINATION
Dec. (daily page)................. d ....
d Corr^n ....(yellow table)
DECLINATION ....

SOLUTION BY H.O. 249
(Assumed Position Method)

Dec. Difference ....
Dec. same name □ or contrary name □ to Lat.
note sign of d in H.O. 249

LHA ....}
Assumed Lat. ....} H.O. 249

→ Tabulated H(Alt.) .... d..... Z ....
d (Dec. Diff.) Corr. ....

H_o ....
H_c .... ← H_c .... Z_n ....
H_o−H_c .... (If H_o is the greater, H_o−H_c is +; + is toward)

DIRECTIONS:

1. Choose assumed long., close to D.R. Long., so that L.H.A. becomes a whole number of degrees.

2. Choose assumed lat. equal to nearest whole degree to D.R. Lat.

3. N.Hem. For L.H.A. between 0° and 180°: $Z_n = 360° - Z$
   For L.H.A. between 180° and 360°: $Z_n = Z$

   S.Hem. For L.H.A. between 0° and 180°: $Z_n = 180° + Z$
   For L.H.A. between 180° and 360°: $Z_n = 180° - Z$

4. (a) Plot assumed lat. and long.

   (b) Lay down bearing of observed body (Z_n).

   (c) From assumed position plot intercept, H_o-H_c, toward or away from bearing of body, according to whether H_o-H_c is + or -.

   (d) Through point found under Step (c) draw a line of position at right angles to the bearing Z_n.

STAR FIX (H.O. 249)
(Vol. 1)

STAR _____   STAR _____   STAR _____

REDUCTION OF SEXTANT OBSERVATION

Ht. of Eye ..... ft.

+	−
......	......

+	−
......	......

+	−
......	......

Index Corr.

Ht. of Eye Corr. ...... (cover table)

3rd. Corr. to H_s ...... table)

Sums

H_s .... H_s .... H_s ....

Total Corr. .... Add if + Sub. if −

H_o .... H_o .... H_o ....

TO FIND GMT

Zone Watch Time .... { may be omitted

Z.D. ....

GMT Watch Time ....

GMT Watch Error .... Sub if fast Add if slow

GMT ....

Greenwich Date .... 19...

GHA ♈ ...hrs .... (daily page)

Corr ...m ... s .... (yellow table)

GHA ♈ ....

Assumed Long. .... Sub if W Add if E

LHA ♈ .... O.O.O'

TO FIND GMT

Zone Watch Time ....

Z.D. ....

GMT Watch Time ....

GMT Watch Error ....

GMT ....

Greenwich Date ....

GHA ♈ ...hrs ....

Corr ...m ... s ....

GHA ♈ ....

Assumed Long. ....

LHA ♈ .... O.O.O'

TO FIND GMT

Zone Watch Time ....

Z.D. ....

GMT Watch Time ....

GMT Watch Error ....

GMT ....

Greenwich Date ....

GHA ♈ ...hrs ....

Corr ...m ... s ....

GHA ♈ ....

Assumed Long. ....

LHA ♈ .... O.O.O'

Assumed Lat. ....   Ass. Lat. ....   Ass. Lat. ....

Repeat H_o ....   Repeat H_o ....   Repeat H_o ....
(from above)   (from above)   (from above)

H_c (H.O.249) .... H_c (H.O.249) .... H_c (H.O.249) ....

$H_o - H_c$ .... $H_o - H_c$ .... $H_o - H_c$ ....

Z_n (H.O.249) .... Z_n (H.O.249) .... Z_n (H.O.249) ....

DIRECTIONS FOR PLOTTING:

(a) Plot assumed lat. and long.

(b) Lay down bearing of observed body (Z_n).

(c) From assumed position plot intercept, $H_0 - H_c$, toward or away from bearing of body, according to whether $H_0 - H_c$ is + or −.

   (If H_0 is the greater, $H_0 - H_c$ is +).

(d) Through point found under Step (c) draw a line of position at right angles to the bearing Z_n.

(e) Apply PRECESSION & NUTATION CORR, if necessary, to FIX or to L.O.P.'s.